During the past decade, mathematics education has changed rapidly, giving rise to a polarization of opinions among the community of research mathematicians. What is the appropriate balance between theory, technique, and applications? What is the role of technology? How do we fulfill the needs of students entering other fields?

This book is the outcome of an effort to create a dialogue about mathematics education that included mathematicians with a wide variety of views. In a conference held at the Mathematical Sciences Research Institute in Berkeley in 1996, more than one hundred mathematics instructors, all dedicated to effective mathematics teaching, engaged in meaningful and sometimes spirited discussions about how mathematics can and should be taught. The conference consisted of formal presentations, question and answer sessions, and informal working groups. This volume contains reports and position papers stemming from all these activities.

Part I deals with general issues in university mathematics education; Part II presents case studies on particular projects; Part III presents a range of opinions on mathematics education in elementary and secondary schools; and Part IV presents the reports of the working groups.

Mathematical Sciences Research Institute
Publications

36

Contemporary Issues in Mathematics Education

Mathematical Sciences Research Institute Publications

Volumes 1–4 and 6–27 are available from Springer-Verlag

Contemporary Issues in Mathematics Education

Edited by

Estela A. Gavosto

University of Kansas, Lawrence

Steven G. Krantz

Washington University in St. Louis, Missouri

William McCallum

University of Arizona, Tucson

CAMBRIDGE UNIVERSITY PRESS

Estela A. Gavosto
Department of Mathematics
University of Kansas
Lawrence, KS 66045
United States

Steven G. Krantz
Department of Mathematics
Campus Box 1146
Washington University in St. Louis
St. Louis, MO 63130
United States

William McCallum
Department of Mathematics
University of Arizona
Tucson, AZ 85721
United States

Mathematical Sciences
  Research Institute
1000 Centennial Drive
Berkeley, CA 94720
United States

Series Editor
Silvio Levy

MSRI Editorial Committee
Hugo Rossi (chair)
Alexandre Chorin
Silvio Levy
Jill Mesirov
Robert Osserman
Peter Sarnak

The Mathematical Sciences Research Institute wishes to acknowledge
support by the National Science Foundation.

Published by the Press Syndicate of the University of Cambridge
The Pitt Building, Trumpington Street, Cambridge CB2 1RP, UK
40 West 20th Street, New York, NY 10011–4211, USA
10 Stamford Road, Oakleigh, Melbourne 3166, Australia

First published 1999

Printed in the United States of America

Library of Congress cataloguing-in-publication data is available.

A catalogue record for this book is available from the British Library.

ISBN 0 521 65255 3 hardback
ISBN 0 521 65471 8 paperback

Writings from the Conference on the Future of Mathematics Education at Research Universities, held at MSRI on December 5 and 6, 1996, and supported in part by National Science Foundation grant DUE 9252521.

Contents

Contemporary Issues in Mathematics Education
MSRI Publications
Volume **36**, 1999

Preface

In the collective unconscious of most working mathematicians, there is the lambent vision of a mathematics department in which mathematicians sit quietly in their offices and prove theorems. Occasionally they are interrupted by a duty to go teach, or to serve on a committee. But, for the most part, they think about mathematics. The mathematics department in this vision has a truce with the University administration: all the mathematicians have research grants, are regularly invited to other universities to give colloquia and to consult, and are recognized scholars; as a result, the administration does not press too hard about the quality of teaching, or how well the department serves the needs of students from other departments in science and engineering.

For better or for worse, the mathematics department described in the last paragraph no longer exists. There are several reasons for this change. Forty years ago, when that chimeric mathematics department *did* exist, it could safely be said that mathematics was an elitist subject. Society's demand for the mathematically fluent was quite small, and those few who survived the mathematics curriculum were more than sufficient to fill the need for teachers, researchers, and mathematical scientists. There was no need for self-examination because what we did seemed to work.

There is now a broad perception that what we do does *not* work. Society now demands a technologically literate work force, and the elitist teaching methodology developed by earlier generations of mathematicians is no longer adequate to the job. More precisely, it is no longer adequate in view of the broad cross-section of society that we are trying to educate, and in view of the level of technical competence that is demanded of those whom we graduate.

Put a different way, we mathematicians must learn to be consciously aware that most of the students in most of our classes are not future mathematicians. With few exceptions, they will not be going on for graduate work in mathematics; and in many cases they may never take an upper division (or a rigorous) mathematics course. We need to understand and embrace the observed fact that a business student has a real need to understand the meaning of the statement "the rate of decrease of inflation is increasing"; this student probably does *not* have any use for knowing how a predator-prey problem can be modeled using a pair of coupled ordinary differential equations.

Forty years ago the observation that 60% or more of freshman fail calculus would have been considered confirmation of the rigors of our curriculum. Now the same statistic (this statistic, and worse, is valid country-wide) signals that something is fundamentally wrong with the way that we teach calculus, and perhaps with the way that we teach undergraduate mathematics overall.

In 1986, Ronald Douglas convened a small conference at Tulane University to examine the state of modern mathematics teaching. By way of this meeting, the "reform" movement in mathematics teaching was born. Although "reform" has a different taste in different people's mouths, it is safe to say that the characteristics of reform are an emphasis on concepts rather than calculations, a creative use of technology, and a stress on geometric insights. Reform teachers use group work, Socratic dialogue, discovery, computers, and other non-traditional methods to get students engaged in the learning process.

Not all mathematics instructors are receptive to the tenets of reform. Many who have taught for decades using lectures, and who feel that their lectures have been effective, are resistive to "throwing out the baby with the bath water" as they are pushed to embrace the new techniques. Debate over the merits of reform, and how it should be implemented, has sometimes been quite heated. In the ensuing discussions, invective has replaced careful reasoning, intuition and conjecture have replace fact and study, and our hard-won scholarly method has often been forgotten.

While many "calculus reform" projects have prospered with the aid of federal grants, it has been convenient for most mathematicians at research universities to ignore the studies and results of the reform movement. Those mathematicians prefer to live in the dream mathematics department described in the first paragraph, and to treat the reform movement as background noise.

The present book is the outcome of an effort to create a dialogue about mathematics education that went beyond calculus reform, and that included mathematicians with a wide variety of views. In a conference held at the Mathematical Sciences Research Institute in Berkeley on December 5 and 6, 1996, with support from MSRI and the National Science Foundation, more than one hundred mathematics instructors, all dedicated to effective mathematics teaching but by different means, engaged in meaningful and sometimes spirited discussions about how mathematics can and should be taught. The conference consisted of formal presentations, question and answer sessions, and more informal "work groups". Our volume contains reports and position papers stemming from all these activities.

As the reader may imagine, the participants in our conference were anxious to be open-minded and to engage in a civilized give-and-take with participants of all stripes. But there was some animated debate, and even some heated dialogue. One of the most interesting interchanges concerned the issue of how "non-proofs" should be presented to a calculus class. For example, if you state the Fundamental Theorem of Calculus and explain why it is true with the usual picture depicting the area under the function, from x to $x + \Delta x$, approximated by the area of a rectangle, then should you take pains to tell the students that this really is not a proof—it is in fact a heuristic? Some traditionalists strongly favored the notion that proofs are sacred, and when you show a class something that is not strictly a proof then you should say so. Others, including some reformists, felt that if you say "here is an idea of why this is true" or "this picture will help you see why this is true," you will have covered all bases and will not have lied. Of course, there is no one correct answer to this issue, and there is much room for disagreement.

The methods that many of us have adopted in our teaching have primarily evolved through repeated experience and through trial-and-error. Few of us have ever had any formal instruction at teaching, and few of us have ever engaged in any formal discussion of issues of pedantry. Thus, for many participants in this conference, there was the joy of discovery of a new sort of discourse. There was also the joy of discovery of new and untapped emotions.

The pedantic issue (about proofs) raised in the last paragraph but one was never settled at our conference. It is safe to say that the main accomplishment of the meeting, apart from giving people the opportunity to make new acquaintances and engage in networking, was to sensitize everyone to a number of important teaching issues and techniques.

And that is really what showing people how to teach is all about: it is decidedly *not* to inculcate in them any particular set of values, nor a particular teaching methodology. Rather, it is to acquaint them with the goals of teaching, the problems that may arise, and with various methodologies that one might use to handle them.

The essays in this volume address the new teaching environment in which we live and work, the newly structured society that we serve, and the new sets of goals and values that are being set for every mathematics department in the country. We hope that they will be of value to everyone who is striving to be an effective teacher. And we also hope that they will be the basis for further productive discussions of teaching issues.

Estela A. Gavosto
Steven G. Krantz
William McCallum

Contemporary Issues in Mathematics Education
MSRI Publications
Volume **36**, 1999

Keynote Address: Mathematicians as Educators

HYMAN BASS

The mathematical sciences professions are in a phase transition, from which they may well emerge smaller, and/or redistributed and much more dispersed. We are not an endangered species, but our health depends on being able to transcend our historic tendencies toward insularity and on our outreach to all of our sister and client communities. This message, in diverse forms, is widely heard today.

The internal mathematical culture continues its deep investigation of the fundamental structures of number, space, dynamics, now with the added exploratory and processing power of new technology. These investigations are guided partly by purely intellectual evolution, but largely also by the natural sciences, to which mathematics furnishes the language and concepts for description, analysis, and modeling. In addition, mathematics provides design and simulation tools for engineering, technology, and for the organization and decision processes of industry. These diverse functions of mathematical thinking and tools are increasingly manifest in many professions, and across the technical workforce.

The phase transition mentioned above involves many partial shifts of focus — from core mathematics toward applications and toward interdisciplinary work with the natural and social sciences, from academic to industrial and laboratory settings, from individual self-directed work to collaborative and multidisciplinary effort, from technical communication with co-specialists to translational communication across disciplinary and cultural boundaries.

Mathematics education is meant to provide appropriate mathematical knowledge, understanding, and skills to diverse student populations. At the post secondary level, such education is entrusted to two large communities. One is based in our system of two-year and community colleges. The other consists of academic mathematical scientists, most of whom have been trained principally to do mathematical research, but for whom the economic base of their profession

This keynote address to the Conference first appeared in the *Notices of the American Mathematical Society* **44**:1 (January 1997), pp. 18–21. It is reprinted here with permission of the publishers. I have greatly benefited from discussion of these ideas with Deborah Ball, Joan Ferrini-Mundy, Hugo Rossi and Lynn Steen.

is now predominantly this educational mission. There has also been a small but distinguished group of scholars doing research and curriculum development in post-secondary mathematics, in the tradition of Pólya — for example Ed Dubinsky, Joan Ferrini-Mundy, Steve Monk, and Alan Schoenfeld.

The shifts described above are reflected in corresponding profound changes in the role of mathematics education. In the post WWII years we had designed a powerful educational model for producing an elite cadre of highly trained and motivated students destined for sophisticated scientific and technical careers. Some very able and committed mathematicians turned their professional energy to this educational task, often with inspiring success. But for the most part the pedagogy was formal, didactic, sometimes brilliant, and often severe. The many students whom it alienated and whom it filtered out of advanced mathematical study were deemed to fail the high standards of our calling. They were seen as lacking "the right stuff". Since the country did not require vast numbers of mathematically trained professionals, and there was sufficient mathematical talent and motivation to survive any pedagogy, this filtering system was considered benign. Many even considered it desirable.

The emergence of a highly competitive and technological world economy has fundamentally enlarged the demands on mathematics education. We now seek, for the broad workforce, levels of scientific and technical competence and literacy that approach what was formerly deemed appropriate for only a select and specialized student population. These same changes make increased demands of technical literacy for responsible and informed participation in our modern democratic society. These pressures give an added practical edge to the traditional argument for the cultural enrichment and intellectual empowerment that mathematical ideas and thinking can confer. When large numbers of students fail and/or leave mathematical study, which is the gateway to such competence and literacy, this is judged now to be the failure — not of the students — but of the educational system. Moreover, the students lost are disproportionately from the minority and female populations that constitute the major influx into the workforce.

The time has come for mathematical scientists to reconsider their role as educators. We constitute a profession that prides itself on professionalism, on an ethos of quality performance and rigorous accountability. Yet academic mathematical scientists, who typically spend at least half of their professional lives teaching, receive virtually no professional preparation or development as educators, apart from the role models of their mentors. Imagine learning to sing arias simply attending operas; learning to cook by eating; learning to write by reading. Much of the art of teaching — the thinking, the dynamic observations and judgments of an accomplished teacher — is invisible to the outside observer. And, in any case, most academic mathematical scientists rarely have occasion to observe really good undergraduate teaching.

While one does not learn good cooking by eating, neither does one learn it just by reading cook books or listening to lectures. Cooking is best learned by cooking, with the mentorship of an accomplished cook; that is, by an apprenticeship model. In fact, teacher education also is designed with a mixture of didactic and apprenticeship instruction. Professional development of academic mathematical scientists as teachers should perhaps be similarly modeled on learning in the context of practice, with only relatively small doses of the more formalized styles of learning with which we are most familiar. Good designs for doing this in a systematic way are not now common. Education professionals can help us in creating and experimenting with such designs.

Effective teaching requires that a teacher know his/her students, to be able not only to explain things to them, but to be able to listen to them, closely, and with understanding. And knowing something for oneself, or for communication to an expert colleague, is not the same as knowing it for explanation to a student. Further, the experience of a mathematical scientist as a learner may not be the best model for the learning of his/her student. These are the kinds of skills and awarenesses that professional development can help cultivate.

Of course there have always been in our professional ranks some very effective, even inspiring, teachers. They have become so through a combination of talent, personal commitment, hard work and practice — and without recourse to professional educators. But do these isolated individuals constitute a model for the educational responsibility of our profession? Are we — and the public we serve — to be content with the condition that some few among our ranks have chosen to take the individual initiative to develop their teaching skills? Imagine, by analogy, abandoning our disciplined education in rigorous mathematics for future researchers to a *laissez faire* system of individual self-instruction, when the impulse happens to be present. How might that affect the quality of our research community?

The disposition of many mathematicians toward the problems of education well reflects their professional culture, which implicitly demeans the importance and substance of pedagogy. Mathematical scientists typically address educational issues exclusively in terms of subject matter content and technical skills, with the "solution" taking the form of new curriculum materials. Curriculum is, indeed, a crucial aspect of the problem, and one to which mathematically trained professionals have a great deal of value to offer. But taken alone it can, and often does, ignore issues of cognition and learning, of multiple strategies for active engagement of students with the mathematics, and of assessing their learning and understanding. Ironically, the mathematical preparation of school teachers is frequently entrusted to these same mathematical scientists, who are often neither trained in nor sensitive to the pedagogical aspects of teaching mathematics to young students. Pedagogy is not something to be added, after the fact, to content. Pedagogy and content are inextricably interwoven in effective teaching.

Pedagogy, like language itself, can either liberate or imprison ideas, inspire or suffocate constructive thinking.

In fact, change on this front has already begun to occur, most notably that stimulated by the so called calculus reform movement. (For an excellent report on this development, see "Assessing Calculus Reform Efforts" (ACRE), by Alan Tucker, MAA, 1995.) To the often skeptical mathematicians outside this activity, the phenomenon is seen as one producing new curricular materials, and of introducing much more systematic uses of technology in teaching calculus. These new materials have been the subject of animated and healthy debate, though some of the opponents have been so stridently and indiscriminately critical as to polarize discussion and impede rational discourse. On the other hand, the people actually engaged in reform calculus teaching typically have a different sense of its significance. They show the same healthy skepticism toward curriculum materials that mathematical scientists have always shown, and they exercise appropriate professional judgment on the manner and extent of use of these materials. What they find most significant about the reform is their personal transformation and the change in their professional practice as teachers. They gain a sense of having become members of a community for which the practice of teaching has become a part of professional consciousness and collegial communication, not unlike their professional practice of mathematics itself. It is the creation of this substantial community of professional mathematician-educators that is, to my mind, the most significant (and perhaps least anticipated) product of the calculus reform movement. This is an achievement of which our community can be justly proud, and which deserves to be nurtured and enhanced. In addition to the ACRE report cited above, the JPBM study on Rewards and Recognition in the Mathematical Sciences is an important gesture in this direction, one that is widely appreciated and cited by our colleagues in other disciplines.

Some might be inclined to cite the calculus reform movement as a case of teaching improvement without the aid of professional educators. On the contrary, there were instances of significant consultation with education specialists. Moreover, the mathematicians who were fully engaged in calculus reform from the early stages, and who had to design programs to prepare the teaching staffs for these new courses, effectively became education specialists with a particular kind of professional expertise. They were funded, and devoted a major part of their time to this development. (I do not rule out the possibility that an education professional may also be a mathematician.) Furthermore, it is quite evident that the pedagogical philosophy that guided the calculus reform powerfully reflected that expressed in the K–12 reform efforts, which came from the thinking of the professional education community.

Once persuaded of the need for improved professional development as teachers, as many mathematical scientists and/or departments (often under external pressure) have become, how do they go about achieving these new goals? How, without prior professional development as teachers, can we mathematical scientists design courses and/or programs to provide this now, for present and future faculty? Part of the answer is that we cannot do it alone, either as mathematical scientists isolated from experienced professional educators (who may themselves also be mathematically trained), or as individual mathematical scientists without the collective support of our ambient departments and institutional environments. Many mathematical scientists have tended to look upon education professionals with doubts bordering on ill-disguised contempt; it is not an easy proposition that we now have much to learn from them, and need their professional help. Much remains to be done to establish contexts for respectful communication and professional collaboration between mathematical scientists and education professionals — from school teachers to people doing education research. This is ultimately a two-way street, along which mathematical scientists can contribute to the disciplinary strengthening of school programs and teaching practice, while the teacher and education research communities can elevate the pedagogical consciousness and competence of academic mathematical scientists.

Mathematics education, unlike mathematics itself, is not an exact science; it is much more empirical, and inherently multidisciplinary. Its aims are not intellectual closure, but helping other human beings, with all of the uncertainty and tentativeness that that entails. It is a social science, with its own standards for evidence, methods of argumentation and theory building, professional discourse, etc. It has an established research base, from which a great deal has been learned in the past few decades; this new pool of knowledge has an important bearing on the educational performance for which academic mathematicians are responsible.

What kinds of things need to be done? At the very least, our graduate students, who regularly perform as TA's or instructors, must be given serious teaching preparation; not only for their duties while graduate students, but also for their roles as possible future university or college faculty, or even as school teachers. Even if their career paths do not take them into the academic world, much of what they need to learn in the way of teaching skills forms part of the broader need for better communications skills in diverse settings. Skill at speaking, instructing, and interacting will make our students better and more effective spokespersons in their work and communities for the importance of sound mathematics education. These skills will also make them more professionally versatile, and more effective in their work. Indeed, such professional development is appropriate for current mathematics faculty as well as for graduate students. In addition, mathematics education provides an important option in the design of new Professional Masters Degree programs in mathematical sciences departments. The resources that support such programs should also provide for ongoing professional educational development of current faculty.

A further important challenge is the design, by mathematical scientists in collaboration with education professionals, of mathematics courses, based in mathematics departments, and devoted to the mathematical preparation of future school teachers. Of course one must distinguish here the needs of elementary teachers from those of secondary teachers. Teacher preparation is an area desperately in need of thoughtful development and experimentation, and which has not received the quality attention by mathematical scientists that it deserves. It invites the possibility of some novel and creative collaborations, where conventional ways of thinking have repeatedly failed to produce desired results.

The above kinds of efforts can be greatly facilitated by networking with colleagues on other campuses at which similar efforts are more highly evolved. There are various activities organized by the Mathematics Education Reform (MER) network and in special sessions at the winter joint meeting of MAA/AMS that support such networking.

While mathematics and mathematics education in the US at the school, college and graduate levels have historically been culturally and professionally separated — a separation visible in the distinct agendas and cultures of the AMS, MAA, AMATYC and NCTM — it becomes clear to anyone who contemplates the needs for improvement of mathematics education in America that this problem cannot be realistically segmented into components for which these four communities take separate and uncoordinated responsibility. As mathematical scientists, as mathematics education researchers, and as teachers in universities, colleges, community colleges and schools, we must begin to see our concerns for graduate, undergraduate and K–12 education as parts of an integrated educational enterprise, in which we have to learn to communicate and collaborate across cultural, disciplinary, and institutional borders, just as we are called upon to do in mathematical sciences research.

HYMAN BASS
COLUMBIA UNIVERSITY
DEPARTMENT OF MATHEMATICS
2990 BROADWAY
NEW YORK, NY 10027-0029
UNITED STATES
    hb@math.columbia.edu

Mathematics Education at the University

Contemporary Issues in Mathematics Education
MSRI Publications
Volume **36**, 1999

On the Education of Mathematics Majors

HUNG-HSI WU

Re-examination of Standard Upper-Division Courses

Upper division courses in college are where math majors learn real mathematics. For the first time they get to examine the foundations of algebra, geometry and analysis, come face-to-face with the deductive nature of mathematics on a consistent basis and, most importantly, learn to do serious theorem-proving. For reasons not unlike these, most mathematicians enjoy teaching these courses more than others. While teaching graduate courses may be professionally more satisfying, it also involves more work, and the teaching of lower division courses — calculus and elementary discrete mathematics — is a strenuous exercise in the suppression of one's basic mathematical impulses. By contrast, the teaching of upper division courses involves no more than doing elementary mathematics the *usual* way: abstract definitions can be offered without apology and theorems are proved as a matter of principle. This is something we can all do on automatic pilot.

But have we been on automatic pilot for too long?

Mathematicians approach these courses as a training ground for future mathematicians. Even a casual perusal of the existing textbooks would readily confirm this fact. We look at upper division courses as the first steps of a journey of ten thousand miles: in order to give students a firm foundation for future research, we feed them technicality after technicality. If they do not fully grasp some of the things they are taught, they will when they get to graduate school or, if necessary, a few years after they start their research. Then they will put everything together. In short, we build the undergraduate education program for our majors on the principle of *delayed gratification*. Whatever their misgivings for the time being, students will benefit in the long haul.

This is the abbreviated version of a longer paper [W4] which discusses the same issues from a slightly different perspective, that of training prospective school teachers. I am indebted to Richard Askey, Paul Clopton, Ole Hald, Ken Ross, and Andre Toom for valuable comments.

However, even the most conservative estimate nation-wide would put no more than 20% of these majors as potential graduate students in mathematics. The remaining 80% — the overwhelming majority — look at the last two years of college as the grand finale of their mathematical experience. To them, their mathematical future is *now*. Given this fact, how should we teach them if we have their welfare in mind? We would want them to understand better the many things they were imperfectly taught in school. We also want them to know what mathematics is about and how mathematics is done. In addition, we owe it to them to give them a sample of the best that mathematics has to offer: some of the major ideas and great theorems in the history of mathematics. But an unflinching assessment would show that, at least for this 80%, we have failed at every step. Take for instance the standard one semester course on complex functions. At the end of such a course, it is perfectly feasible to explain the meaning and significance of the Riemann Mapping theorem, the Dirichlet problem, and the Riemann Hypothesis. Embedded in these three topics are ideas that have helped shape the course of mathematics in the past century and a half, and are therefore ideas which would interest these majors. But how many students in complex functions know about this piece of history even if on rare occasions the Riemann Mapping Theorem is stated and proved? Instead, such courses spend the time on the proofs of the general form of the Cauchy theorem and other technical facts. Another conspicuous example is the recent proof of Fermat's Last Theorem. How many of our majors have the vaguest idea that this proof is an achievement of "enormous humanistic importance" (words of Elliott Lieb)? If we are unhappy about the answers to these questions, we have only ourselves to blame. After all, we are the ones who design this guided tour of the edifice we call mathematics, a tour that allows these majors to see only the nuts and bolts in its foundation but never its splendor or even the *raison d'être* of some of its interior designs. We have let them down. In our effort to nurture future mathematicians, which is undoubtedly an essential goal of mathematics education, we have neglected the education of the remaining 80% of our majors who put themselves under our care. We forget that they too are part of our charge.

The usual defense of this philosophy of education would argue that, far from a case of neglect, this system came about by design. For, by giving students a firm technical grounding with ample exposure to abstract theorems and rigorous proofs, we also give them the tools to explore on their own. Eventually, they will acquire the necessary perspective and overall understanding of the details so that they will look back on all they have learned and enlightenment ensues. Or so the theory goes.

This is what I call the Intellectual Trickle-down Theory of Learning: aim the teaching at the best students, and somehow the rest will take care of themselves. In practice, however, most of the students who do not go on to graduate school in mathematics are not among those with a strong enough interest or firm enough mastery of the fundamentals to dig deeper for further understanding.

Consequently, the college education of these students is long on technical details that they cannot digest but short on the minimal essential information that would enable them to understand even the elementary facts from "an advanced standpoint". They go out into the world impoverished in both technique and information for the simple reason that we never had them in mind when we designed our curriculum.

What lends a sense of urgency to this unhappy situation is the presence of future school mathematics teachers among the 80% in question. When they go into the classroom so mathematically ill-equipped, they cannot help but victimize the next generation of students. Some of the latter come back to the university and the vicious circle continues. We pay a high price indeed for our neglect.

This particular aspect of our collective failure to educate the majority of majors bears on the current mathematics education reform in K–14 (i.e., from kindergarten to the sophomore year in college). This reform (cf. [W1, W2, W3]) has by and large ignored the critical issue of the technical inadequacy of school mathematics teachers.[1] Not coincidentally, there is a corresponding de-emphasis of content knowledge in favor of pedagogy in schools of education across the nation [K, NAR]. For example, it was already explicitly pointed out back in 1983 that "Half of the newly appointed mathematics, science, and English teachers are not qualified to teach these subjects..." [NAR]. To this vitally important area of school education, the most direct contribution the mathematical community can make would seem to be to design a better education for prospective school teachers. Our failure to do this was what gave the original impetus to the writing of this paper.[2]

Let me illustrate the various threads of the preceding discussion with a concrete example. In the spring semester of 1996, I taught a course in introductory algebra which was attended by math majors who did not intend to continue with graduate study in mathematics, including a few prospective school teachers. Before I discussed the field of rational numbers, I asked how many of them knew *why* $1/(1/5) = 5$. I waited a long time, but not a single hand was raised while a few shook their heads.[3] One cannot understand the significance of this fact unless one realizes that the subject of fractions is one of the sore spots in school mathematics education. Herb Clemens [C] relates the following story:

> Last August, I began a week of fractions classes at a workshop for elementary teachers with a graph paper explanation of why $\frac{2/7}{1/9} = 2\frac{4}{7}$. The reaction of my audience astounded me. Several of the teachers present were

---

[1] I am referring to the generic case, of course. There are many excellent teachers.

[2] In this context, I am obliged to point out that this paper does not do justice to the complex issue of how best to educate prospective teachers; what it does is merely to suggest ways to give these prospective teachers a better mathematical training *given the existing requirements of a math major*.

[3] My colleague Ole Hald has suggested to me that one explanation of the lack of response could be the lack of a proper context for the students to understand such a question.

simply terrified. None of my protestations about this being a preview, none of my "Don't worry" statements had any effect.

Or, take another statement from the NCTM Standards [NC, p. 96]:

> This is not to suggest, however, that valuable instruction time should be devoted to exercises like $\frac{17}{24} + \frac{5}{18}$ or $5\frac{3}{4} \times 4\frac{1}{4}$, which are much harder to visualize and unlikely to occur in real-life situations.

This suggestion concerning the teaching of fractions occurs in the Standard on Estimation and Computation in Grades 5–8 of [NC]. It is difficult for a mathematician to imagine that students going into high school (9th grade) would have trouble *computing* simple products and sums such as above, but this difficulty evaporates as soon as we look into how fractions are taught in grades 4–8. Take the standard Addison-Wesley Mathematics for Grades 4–8, for example [EI]. There the definition of the addition of fractions unnecessarily brings in the LCM of denominators, and the division of fractions is defined using circular reasoning (for the latter, see p. 232 of Grade 6, p. 182 of Grade 7 and p. 210 of Grade 8), to name just two problems off-hand. Unfortunately, students do not get to learn substantially more about the rational number system in high school because, once there, they take algebra which assumes they know how to compute with rational numbers. Thus by the time students come to the university, very often their understanding of the rationals remains in the primitive state reflected in the two quotations above.

In the light of this glaring weakness in our average math major's mathematical arsenal, let us take a look at what he or she is taught about the rationals in a typical course in introductory algebra. We first introduce the notion of an integral domain D, and construct out of D its quotient field by introducing equivalence classes of ordered pairs $\{(p, q)\}$, show that addition and multiplication among these equivalence classes are well defined and that they form a field F. Then we define the canonical injective homomorphism from D to F, identify D with its image, and explain to the students that henceforth all nonzero elements of D will have multiplicative inverses in F. After all that, we will mention that if we replace D by the integers, then the F above would be the field of rational numbers. Whether or not one would explicitly point out the relationship between the common assertion $1/(1/b) = b$ and the general fact that $(b^{-1})^{-1} = b$ would depend very much on the instructor, and in any case, even if this is done, it would be a passing remark and no more. In the meantime, the average math major is overwhelmed by this onslaught of newly acquired concepts: integral domain, field, equivalence class, injection, and homomorphism. To most beginners, these are things at best half understood. Thus expecting them to come to grips with the rational numbers by way of the concept of a quotient field is clearly no more than a forlorn hope. It therefore comes as no surprise that, in the midst of such uneasiness and uncertainty, the average student would fail to gain any new insight into such fundamentals as $1/(1/b) = b$ or $(-p)(-q) = pq$. Some of them soon go into school classrooms as teachers and, things being what they are, their students too can be counted on to fear such simple tasks as $\frac{17}{24} + \frac{5}{8}$.

Some Proposed Changes

One way to resolve the difficulty concerning the education of our math majors who do not go on to graduate work in mathematics is to teach them in a separate track. There are obstacles that stand in the way of such a simple recommendation, to be sure, but none seems to be more formidable than the suspicion that it is really *infra dig* for a "good" department to offer "watered down" courses to its own majors. Thus even if money is available to teach two tracks, fighting this prejudice will not be easy. The education of over 80% of our majors is however too serious an issue to be glossed over by institutional or professional prejudices. It is time that we meet this problem head on by discussing it in public.

There is perhaps no better explanation of why "different" is not the same as "watered down" than to list a few of what I believe to be the desirable characteristics of courses designed for students who do not pursue graduate work in mathematics.

(1) *Only proofs of truly basic theorems are given, but whatever proofs are given should be complete and rigorous.* On the one hand, we are doing battle with time: given that there are many topics we want the students to be informed about, the proofs of some of our favorite theorems (e.g., the Jordan canonical form of a linear transformation or the implicit function theorem) would have to go in favor of other issues of compelling interest (e.g., historical background or motivation). On the other hand, we also want them to understand that proofs are the underpinning of mathematics. Thus any time we present a proof, we must make sure that it counts.

(2) *In contrast with the normal courses which are relentlessly "forward-looking" (i.e., the far-better-things-to-come in graduate courses), considerable time should be devoted to "looking back".* In other words, there should be an emphasis on shedding light on elementary mathematics from an advanced viewpoint. One example is the cleaning up of the confusion about rational numbers (cf. the discussion in the preceding section). Another is the elucidation of Euclidean geometry and axiomatic systems.

(3) *Keep the course on as concrete a level as possible, and introduce abstractions only when absolutely necessary.* The potency of abstract considerations should not be minimized, but we have to be aware of the point of minimal return, when any more abstraction would decrease rather than promote students' understanding. The example of the construction of a quotient field from an integral domain in the context of rational numbers has already been given above.

(4) *Ample historical background should be provided.* This idea is by now so widely accepted that no argument need be given save to point out that a knowl-

edge of the evolution of a concept or theorem helps to break down students' resistance to abstractions.

(5) *Provide students with some perspective on each subject, including the presentation of surveys of advanced topics.* For example, in discussing diagonalization of matrices, students would benefit from a discussion of the various canonical forms — *without proofs* — because they need to understand that diagonalization is not an isolated trick but a small part of the general attempt to simplify and classify all mathematical objects. Along this line, I cannot help but be amazed by the general tendency in most texts to refrain from discussing any topic that is "out of logical order". The desire to develop the subject *ab initio* is well taken, but what is there to fear about exposing students to interesting advanced ideas without proofs, so long as the advanced nature of the material is made clear to them? The way we learn is hardly "logically-linear"; otherwise there would be no incentive for any of us to go to colloquium lectures. So why deny the students the same opportunity to learn when we can so easily provide it?

(6) *Give motivation at every opportunity.* The usual complaint about school mathematics degenerating into rote-learning is fundamentally a reflection of the fact that most teachers were themselves never exposed to any motivation for every concept, lemma, and theorem that they learned. It is incumbent on the college instructors to break this vicious circle.

As an example of how to implement some of these changes, consider the teaching of *Dedekind cuts* to students in introductory analysis. As is well-known, there are two kinds of Dedekind cuts. If we try to construct $\mathbb{R}$ out of $\mathbb{Q}$, then we would define:

(1) A **real number** is an ordered pair (A, B) of nonempty subsets of $\mathbb{Q}$, so that $A \cup B = \mathbb{Q}$, and $a < b$, $\forall a \in A$, $\forall b \in B$.

On the other hand, if we try to define $\mathbb{R}$ as a complete ordered field, then we postulate (in the original words of Dedekind, 1872):

(2) "If all points of the straight line fall into two classes such that every point of the first class lies to the left of every point of the second class, then there exists one and only one point which produces this division of all points into two classes, this severing of the straight line into two portions" [D, p. 11].

Beginners are usually befuddled by either of these quaint statements. Now suppose in the classroom we have to use (1) to construct $\mathbb{R}$. We would first point out the relationship between (1) and (2), describe the state of the calculus in Dedekind's time, and make students understand that there was a real need for a non-mystical approach to the real numbers. Clearly something had to be done, and Dedekind's was the first successful contribution. Did he get this idea out of the blue? Hardly. Again, in his own words (1887): "...if one regards the irrational number as the ratio of two measurable quantities, then is this manner of determining it already set forth in the clearest possible way in the celebrated definition which Euclid gives of the equality of two ratios (*Elements*, V, 5). This same most ancient conviction has been the source of my theory ... to lay the foundations for the introduction of irrational numbers into arithmetic" [D, pp. 39–40]. Students at this point should be curious about this famous definition in *Elements*, V, 5. Historians agree that this was in fact the creation of Eudoxus (ca. 408–355 B.C.), and it states [E, p. 114]:

"Magnitudes are said to be **in the same ratio**, the first to the second and the third to the fourth, when, if any equimultiples whatever be taken of the first and the third, and any equimultiples whatever of the second and the fourth, the former equimultiples alike exceed, are alike equal to, or alike fall short of the latter equimultiples respectively taken in corresponding order."

Naturally, few can penetrate this kind of prose. This then gives the instructor an excellent opportunity to extol the virtues of the symbolic notation, something that most students take for granted without any appreciation. In symbols, the preceding paragraph merely states, word for word: "$a_1 : a_2 = a_3 : a_4$ if and only if given any positive integers m and n, $ma_1 > na_2$ implies $ma_3 > na_4$, $ma_1 = na_2$ implies $ma_3 = na_4$, and $ma_1 < na_2$ implies $ma_3 < na_4$." One can make this even more accessible by using modern mathematical language: "$a_1/a_2 = a_3/a_4$ if and only if for any rational number n/m, the following conditions hold: $n/m < a_1/a_2$ implies $n/m < a_3/a_4$, $n/m = a_1/a_2$ implies $n/m = a_3/a_4$, and $n/m > a_1/a_2$ implies $n/m > a_3/a_4$. Or, more simply:

"$a_1/a_2 = a_3/a_4$ if and only if the cuts in $\mathbb{Q}$ induced by both a_1/a_2 and a_3/a_4 are equal."

This then brings us to the previous statement (2) of Dedekind.

It remains to explain to students why Eudoxus would even dream of such a tortuous definition of **equal ratio**. Couldn't he just divide the two pairs of real numbers? Now is the time to discuss the Greeks' veneration of rational numbers in the 5th century B.C., the absence of any concept of "real numbers" back then, the subsequent crisis precipitated by the discovery of incommensurable quantities, and Eudoxus' brilliant achievement in defining incommensurable quantities using only the rationals. Dedekind's insight was to realize that Eudoxus' *ad hoc* definition in fact suffices to pin down the real number field.

Such a detour in a beginning course in analysis takes time. If one's aim is to usher students to the frontier of research in the most efficient manner possible, this detour would be ill-advised. But if one instead tries to instill a little understanding together with some mathematical culture in these students before they leave mathematics, the detour would be well worth it. So it becomes important to know whether we are teaching the 20% or the other 80%.

An Experiment

What I will take up next is the description of a small experiment I am currently conducting at Berkeley in an attempt to implement the preceding philosophy of teaching. Starting in the spring semester of 1996, I have been offering an upper

division course each semester specifically for "math majors who do not go on
to graduate school in mathematics". At the time of writing (April 1997), I
have taught introductory algebra and linear algebra, and am currently teaching
differential geometry. The remaining courses I hope to have taught by June of
1999 (in some order) are history of mathematics, introductory analysis, complex
functions and classical geometries.[4] Having taught two such courses and being
in the middle of a third one, I would like to recount my experience in some detail.
Since there is as yet no tradition of teaching upper divisional math courses with
the philosophical orientation propounded here, others might find this account to
be of some value.

Two general observations emerged from this experience. The first one is that
it is very difficult, if not impossible, to find an appropriate text for such a course.
Almost all the standard texts are written to prepare students for graduate work
in mathematics. At the other extreme are a few texts that try to be "user-
friendly" by trivializing the content. Neither would serve my purpose. The
other observation is that in this approach to upper divisional instruction, the
trade-offs are quite pronounced even for someone who is prepared for them. Let
me be more specific by discussing separately the two courses I have taught thus
far.

For the introductory algebra course, the announced goals of the course were
the solutions of the three classical construction problems and the explanation of
why the roots of certain polynomials of degree ≥ 5 cannot be extracted from the
coefficients by use of radicals. In more detail, the first two thirds of the course
were devoted to the following topics:

Part 1: The quadratic closure of $\mathbb{Q}$, constructible numbers, $\mathbb{Z}$ and $\mathbb{Q}$ re-
visited, Euclidean algorithm and prime factorization, congruences modulo
n, fields, polynomials over a field, irreducibility and unique factorization
of polynomials, Eisenstein, complex numbers and the fundamental theorem of algebra, roots of unity and cyclotomic polynomials, field extensions
and their degrees, solutions of the construction problems, constructibility
of regular polygons.

The last third of the course was more descriptive in nature. It consisted of the
following:

Part 2: Isomorphism of fields, automorphisms relative to a ground field,
root fields, computations of automorphisms, groups and basic properties,
solvable groups, Galois group of an extension, radical extensions, the theorems of Abel and Galois.

---

[4]The absence of number theory in this list may come as a surprise to some, but I did not
create a separate number theory course for this track because I cannot imagine there is anything
in a beginning number theory course that should not be known to one and all regardless of
students' needs.

The main reason for the decision to direct the whole course towards these classical problems was the hope that students' prior familiarity with these problems would entice them to learn the abstract algebraic concepts needed for their solutions. My hope was not in vain: a few of my 19 students told me after the course was over that they had worked harder for this course than for other math courses because they could relate to what they were learning. Sadly, with the gutting of Euclidean Geometry and the de-emphasis of purely mathematical questions (without real-world relevance) in the present school mathematics education reform, soon students will go to college not knowing any of these famous problems. This is a potential educational crisis that mathematicians must do all they can to avert.

Another reason for this choice of materials was the historical connection. Since these problems gave impetus to almost all the important advances in algebra up to the nineteenth century, the course provided a natural setting to delve into the historical roots of the subject. I was able to discuss the slow emergence of the symbolic algebraic notation, Cardan, Tartaglia, Fermat, Gauss, Galois and Abel.

Compared with more standard approaches to introductory algebra, some losses and gains are worth noting. Because I consider the topics in Part 1 to be basic to a teacher's understanding of algebra, every theorem there was proved, with the exception (of course) of the fundamental theorem of algebra, the transcendence of π, and the theorem of Gauss on the constructibility of regular polygons. Students received, and welcomed, a careful and rigorous treatment of $\mathbb{Q}$ and $\mathbb{C}$. In addition, because Part 1 gives a detailed treatment of the polynomial ring over $\mathbb{R}$, these students got to understand this important object — important especially for the future teachers — much better than otherwise possible. For example, I made a point of proving for them the technique of partial fractions which is usually imperfectly stated and used for integration in calculus. On the other hand, this choice of topics entails certain glaring omissions: no PID's and no UFD's, no general construction of the quotient field of a domain, and no serious discussion of ideals, ring homomorphisms and quotient rings.

The omissions in Part 2 are much more serious. With only five weeks to treat these sophisticated topics, there was hardly time to prove any theorem other than the simplest. Even at the most basic level, there was no discussion of homomorphism between groups, and hence also no discussion of the relationship between normal subgroups and the kernel of a homomorphism, and theorems about the existence of subgroups of an appropriate order were hardly mentioned. In exchange, students were given exposure to the fantastic ideas of Galois theory — without proofs, to be sure — and the hope is that perhaps one day some of them would revisit the whole terrain on their own.

Implicit in the preceding syllabus is the fact that no noncommutative mathematics appears until two thirds of the way into the course when groups enter the discussion. This is by design. It seems to me that students taking such a course are confronted with proofs in a serious way for the first time, and that is enough of a hurdle without their being simultaneously overloaded by noncommutativity as well. Because it took mankind more than two thousand years after Euclid to face up to noncommutativity, it seems unfair that beginning students are not given a few weeks of reprieve before being saddled with it. The presentation in almost all the standard texts in algebra, beginning with groups and followed by rings and fields, mimics the order adopted by van der Waerden in his pioneering text of 1931 [WA]. However, van der Waerden was writing a research monograph, and there seems to be very little reason why undergraduate texts should follow suit without due regard for pedagogy. Along this line, let it be said that the first American undergraduate textbook on (modern) algebra, *A Survey of Modern Algebra* of Birkhoff and MacLane [BM], does begin with integers, commutative rings, and more than a hundred pages of commutative mathematics before launching into groups. There is a reason why Birkhoff–MacLane is still in print after 55 years. Having said that, I want to reiterate a serious concern in the teaching of such a course, which is the pressing need of an appropriate textbook.

Next, let us turn to linear algebra. Among upper division math courses, this course may be the only one which is as rich as calculus in nontrivial scientific applications. Furthermore, this is also a subject almost tailor-made for computers and therefore the consideration of computational simplicity plays an important role. For the benefit of those students who do not go on to higher mathematics, a course on linear algebra that emphasizes both of these facts in place of topics like the Cayley-Hamilton theorem and the rational canonical form would seem to be more educational.[5] While it is true that standard courses usually pay some attention to computational simplicity at the beginning when discussing Gaussian elimination, it is also true that they quickly drop this consideration the rest of the way. I would prefer that this consideration be the underlying theme of the whole course because, in scientific applications, savings in cost and time are important.[6] The syllabus for the course I gave in the fall of 1996 is then:

> Elementary row operations, Gaussian elimination, existence and uniqueness of LU decomposition of nonsingular square matrices, approximate solutions of ODE's by linear systems, review of vector spaces and associated concepts, LU decomposition in general, the row space, column space and solution space of a general matrix, the rank-nullity theorem, application to electrical networks, inner product spaces, orthogonal projection on a subspace, least square approximations, Gram-Schmidt, the QR decomposition of a nonsingular matrix, signal processing and the fast Fourier transform,

---

[5] Of course students should learn the Cayley-Hamilton theorem and the rational form too *if* there is time.

[6] The issue of roundoff errors was mentioned in my course but not pursued. It seems to me that this would be better handled in a course on numerical analysis.

determinants, eigenvalues and eigenvectors, diagonalization, applications to Fibonacci sequence, symmetric and Hermitian matrices, quadratic forms.

This is clearly a course on matrices rather than abstract vector spaces. Even so, some standard topics on matrices are conspicuous by their absence: the minimal polynomial, Cayley-Hamilton, Jordan canonical form and rational canonical form. But perhaps the most distressing aspect was my inability, due to lack of time, to impress on the students the advantage of knowing the coordinate-free point of view. It goes without saying that there was no mention in the course of invariant subspaces, dual spaces, and induced dual linear transformations. The difficulty with some of these deficiencies could have been alleviated by making reading assignments in an appropriate textbook, one in which the applications are presented with integrity and the abstract point of view is treated with the clarity and precision that befit a mathematics text. However, the scarcity of textbooks suitable for the kind of teaching under discussion continues to call attention to itself.

What remains to be said are the gains that go with such losses, namely, the interesting applications that throw a completely different light on abstract linear algebra. Personally, I must admit to having been enthralled by the applications of the fast Fourier transform to signal processing, and this sense of enchantment would be shared by the students too if the latter is properly explained. No less interesting is the way the QR decomposition helps save time and cost in formulating precise experimental laws using the least squares method. These rather pronounced trade-offs in such an approach to teaching will undoubtedly continue to invite strong reactions from each of us.

Finally, an ongoing concern of my colleagues is that by not proving *every* theorem, such courses run the risk of giving students a distorted perception of the fundamental nature of proofs in mathematics. Whether or not such a danger would be realized in the classroom seems to me to depend very much on the way a course is actually taught. There are no foolproof pedagogical strategies. For the cases at hand, I have appended the final exams of the preceding two courses for the readers' inspection. One can at least get a sense of what was emphasized in these courses even if not of what was actually achieved.

The Summing Up

Upper division courses should enlarge a student's basis of mathematical knowledge. With this in mind, I find it necessary to lecture throughout these courses to ensure that a minimum number of topics be covered. At the same time, I also attach a great deal of importance to homework assignments. To help students with problem solving, which includes not only getting the solutions but also learning how to write intelligible proofs, I either pass out solution sets or schedule extra problem sessions each week. While one hesitates to make an unconditional recommendation of such time-consuming efforts as part of our teaching duty, one should also ask if most students can learn much in these courses by simply attending lectures.

It should be emphasized that what has been proposed above is an *alternative* to the standard courses, not a replacement. If we are still committed to the concept of the university as a repository of knowledge, then we must insure the continuity of this knowledge by producing future mathematicians. The standard upper division courses therefore play a vital role in honoring this commitment. On the other hand, a university is also an educational institution and the education of the majority of the math majors must also be taken seriously. The aim of the present proposal is therefore to suggest a way for the university to better fulfill this dual obligation, and *not* to suggest a radical change in undergraduate education. In practical terms, the suggested alternative can be implemented more easily in large institutions than in small colleges. In the former, scheduling parallel sections of the same course catering to different student clienteles does not present much difficulty. One can only hope that this suggestion, in one form or another, will be discussed in these institutions. In the smaller colleges, any change would seems to come only with some special effort or ingenuity.

Finally, the present proposal is very germane to the current education reform in K–14. There is a tendency in this reform to make sweeping changes: wholesale replacements of existing curricula or pedagogies are routinely recommended. The idea that one can offer alternatives to *some* but not all of what are already in place appears to be foreign to the reformers, as is the need to clearly delineate the liabilities as well as virtues of every one of the proposed changes. Education is a multi-faceted enterprise which, for better or for worse, involve both politicians and the public. We accept the fact that the latter two thrive in the world of hyperbole. However, most of the reform measures have been proposed in the academic community. I hope I will be forgiven for my temerity if I offer the reminder that, insofar as education is still an academic subject, the academics who propose education reform should make an effort to conform to the minimal standards of intellectual integrity and candor. Had such an effort been made, much of the rancor of the present reform would have disappeared and a more rational course of action would have resulted. For the sake of the next generation — and the reform is nothing if not about the welfare of the next generation — let us restore such integrity and candor to our discussions.

Appendix: Sample Final Exams

Introduction to Abstract Algebra (Math 113)
FINAL EXAM
May 13, 1996   8–11 am   Prof. Wu

1. (5%) Prove that for an integer n, $3 \mid n \Longleftrightarrow 3 \mid$ (sum of digits of n).

2. (5%) Let $f(x) = x^n + a_{n-1}x^{n-1} + \cdots a_1 x + a_0$ be a polynomial with integer coefficients, and let r be a rational number such that $f(r) = 0$. Show that r has to be an integer and $r \mid a_0$.

3. (5%) Find a minimal polynomial of $\sqrt[3]{1 + \sqrt{3}}$ over $\mathbb{Q}$. (Be sure to prove that it is minimal.)

4. (5%) Let n be a positive integer ≥ 2 such that $n \mid (b^{n-1} - 1)$ for all integers b which are not a multiple of n. What can you say about n?

5. (5%) Do the nonzero elements of $\mathbb{Z}_{13}$ form a cyclic group under multiplication? Give reasons.

6. (10%) Let p be a prime.

(a) Prove: $p \mid \binom{p}{k}$ for $k = 1, \ldots p-1$, where $\binom{p}{k} \equiv \dfrac{p!}{k!(p-k)!}$.

(b) Prove: the mapping $f : \mathbb{Z}_p \to \mathbb{Z}_p$ defined by $f(k) = k^p$ for all $k \in \mathbb{Z}_p$ is a field isomorphism.

7. (10%) Is $x^4 + 2x + 3$ irreducible over $\mathbb{R}$? Is it irreducible over $\mathbb{Q}$? Give reasons.

8. (10%) Let $F \equiv \{a + ib : a, b \in \%Q\}$ and let $K \equiv \mathbb{Q}[x]/(x^2 + 1)\mathbb{Q}[x]$. Show that F is isomorphic to K as fields by defining a map $\varphi : F \to K$ and show that φ has all the requisite properties.

9. (10%) If β is a root of $x^3 - x + 1$, find some $p(x) \in \mathbb{Q}[x]$ so that $(\beta^2 - 2)\,p(\beta) = 1$.

10. (10%) Let $\zeta = e^{i2\pi/3}$. Compute $(\mathbb{Q}(\zeta, \sqrt[3]{5}) : \mathbb{Q}(\zeta))$.

11. (25%) (In (a)–(d) below, each part could be done independently.)

(a) Assume that if p is a prime, then $x^{p-1} + x^{p-2} + \cdots + 1$ is irreducible over $\mathbb{Q}$. Compute $(\mathbb{Q}(\cos(2\pi/7) + i\sin(2\pi/7)) : \mathbb{Q})$. (Each step should be clearly explained.)

(b) Suppose the regular 7-gon can be constructed with straightedge and compass. Explain why $(\mathbb{Q}(\cos(2\pi/7) : \mathbb{Q}) = 2^k$ for some $k \in \mathbb{Z}^+$.

(c) If $F \equiv \mathbb{Q}(\cos(2\pi/7))$, show that $(F(i\sin(2\pi/7)) : F) = 1$ or 2.

(d) Use (b) and (c) to conclude that if the regular 7-gon can be constructed with straightedge and compass, then $(\mathbb{Q}(\cos(2\pi/7) + i\sin(2\pi/7)) : \mathbb{Q}) = 2^m$ for some $m \in \mathbb{Z}^+$.

(e) What can you conclude from (a) and (d)? What is your guess concerning the construction of the regular 11-gon, the regular 13-gon, the regular 23-gon, etc.?

Linear Algebra (Math 110)
FINAL EXAM
Dec 11, 1996   12:30–3:30 pm   Prof. Wu

1. (5%) Find the determinant of $\begin{bmatrix} 2 & 2 & 0 & 4 \\ 3 & 3 & 2 & 2 \\ 0 & 1 & 3 & 2 \\ 2 & 0 & 2 & 1 \end{bmatrix}$ and show all your steps.

2. (15%) Let $A = \begin{bmatrix} 1 & -1 & -1 \\ -1 & 1 & -1 \\ -1 & -1 & 1 \end{bmatrix}$. Find its eigenvalues and the corresponding eigenvectors. Also find a 3×3 matrix S and a diagonal matrix D so that $S^{-1}AS = D$.

3. (5%) If Q is an $n \times n$ orthogonal matrix, what is $\det Q$? What are the eigenvalues of Q? Give reasons.

4. (15%) Let F_k denote the Fourier matrix of dimension k. Define for each n:

$$Y_{2n} = \begin{bmatrix} y_0 \\ y_1 \\ y_2 \\ \vdots \\ y_{2n-1} \end{bmatrix}, \qquad Y_{\text{odd}} = \begin{bmatrix} y_1 \\ y_3 \\ y_5 \\ \vdots \\ y_{2n-1} \end{bmatrix}, \qquad Y_{\text{even}} = \begin{bmatrix} y_0 \\ y_2 \\ y_4 \\ \vdots \\ y_{2n-2} \end{bmatrix}$$

Also let $w = e^{i\frac{2\pi}{2n}}$ and $W = \begin{bmatrix} 1 & & & & \\ & w & & & \\ & & w^2 & & \\ & & & \ddots & \\ & & & & w^{n-1} \end{bmatrix}$. Then the fast Fourier transform can be described as follows:

$$F_{2n}Y_{2n} = \begin{bmatrix} F_n Y_{\text{even}} + W F_n Y_{\text{odd}} \\ F_n Y_{\text{even}} - W F_n Y_{\text{odd}} \end{bmatrix}$$

Now let $\rho(2n)$ denote the minimum number of operations needed to compute $F_{2n}Y_{2n}$. Prove: $\rho(2^k) \leq k2^k$ for every integer $k \geq 1$. (Recall: an "operation" means either a multiplication-and-an-addition or a division.)

5. (10%) We want a plane $y = C + Dt + Ez$ in $y - t - z$ space that "best fits" (in the sense of least squares) the following data: $y = 3$ when $t = 1$ and $z = 1$; $y = 5$ when $t = 2$ and $z = 1$; $y = 6$ when $t = 0$ and $z = 3$; and $y = 0$ when $t = 0$ and $z = 0$. Set up, but do not solve the 3×3 linear system of equations that the least squares solution C, D, E must satisfy.

6. (10%) Let $v_1, \ldots, v_k$ be eigenvectors of an $n \times n$ matrix A corresponding to *distinct* eigenvalues $\lambda_1, \ldots, \lambda_k$, respectively, where $k \leq n$. Prove that $v_1, \ldots, v_k$ are linearly independent.

7. (10%) Suppose a real $n \times n$ matrix A has n distinct real eigenvalues. Is there necessarily a *real* $n \times n$ matrix S so that $S^{-1}AS$ is diagonal? Explain.

8. (10%) If the eigenvalues of A are $\lambda_1, \ldots, \lambda_n$ (not necessarily distinct), what are the eigenvalues of A^k where k is an integer ≥ 1? If A is nonsingular, what are the eigenvalues of A^{-k} for $k \geq 1$? Give reason.

9. (15%) Let P be the projection matrix which projects $\mathbb{R}^n$ onto a k-dimensional subspace $W \subset \mathbb{R}^n$, where $0 < k < n$. Enumerate all the eigenvalues of P and for each eigenvalue, describe all its eigenvectors.

10. (5%) Let A be an $n \times n$ matrix and let A' be obtained from A be Gaussian elimination. Do A and A' necessarily have the same eigenvalues? Give reasons.

References

[BM]   Garrett Birkhoff and Saunders MacLane, *A Survey of Modern Algebra*, Macmillan, New York, 1941.

[C]   C. H. Clemens, "Can university math people contribute significantly to precollege mathematics education (beyond giving future teachers a few preservice courses?)", *CBMS Issues in Mathematics Education* **5**, Amer. Math. Soc., 1995, 55–59.

[D]   Richard Dedekind, *Essays in the Theory of Numbers*, Dover, New York, 1963.

[EI]   Robert Eicholz et al., *Addison-Wesley Mathematics* for Grades 4–8, Addison-Wesley, Menlo Park, CA, 1995.

[EU]   Euclid, *The Thirteen Books of the Elements*, Volume 2, Dover, New York, 1956.

[K]   Rita Kramer, *Ed School Follies*, The Free Press, 1991.

[NAR]   *A Nation at Risk*, U.S. Department of Education, Washington D.C., 1983.

[NC]   *Curriculum and Evaluation Standards for School Mathematics*, National Council of Teachers of Mathematics, Reston, 1989. Available at http://www.enc.org/online/NCTM/280dtoc1.html.

[WA]   B. L. van der Waerden, *Moderne Algebra*, J. Springer, Berlin, 1930–31.

[W1]   H. Wu, "What about the top 20%?" Manuscript of a talk given at the Annual Meeting of the Amer. Math. Soc., Jan. 4, 1995; available from the author.

[W2]   H. Wu, "The mathematics education reform: why you should be concerned and what you can do", Amer. Math. Monthly **104**:12 (December 1997).

[W3]   H. Wu, "The mathematics education reform: What is it and why should you care?", 1997. Available at http://math.berkeley.edu/~wu.

[W4]   H. Wu, "On the training of mathematics teachers", 1996. Available at http://math.berkeley.edu/~wu.

HUNG-HSI WU
UNIVERSITY OF CALIFORNIA BERKELEY
DEPARTMENT OF MATHEMATICS
BERKELEY, CA 94720-3840 UNITED STATES
   wu@math.berkeley.edu

Contemporary Issues in Mathematics Education
MSRI Publications
Volume **36**, 1999

The Mathematics Major at Research Universities

PETER G. HINMAN AND B. ALAN TAYLOR

In the past decade the focus in university science and mathematics education has shifted from providing an adequate supply of world-class professional
scientists to the broader agenda of providing excellent education in science and
mathematics to all undergraduate students. In the words of David Goodstein,
"...the United States has, simultaneously and paradoxically, both the best scientists and the most scientifically illiterate young people: America's educational
system is designed to produce precisely that result. America leads the world
in science — and yet 95 percent of the American public is scientifically illiterate." In an environment where jobs that provide decent economic opportunity
demand skills far more sophisticated than those required in the past, universities
are now being called upon to provide all of their students both with a supportive
environment for acquiring these skills and with the ability to continue learning
throughout a lifetime in the workforce.

Many of the current educational changes are driven by this new agenda. Major efforts have been and are being made to improve learning, primarily at the
freshman-sophomore level, for students who are not planning to become mathematicians, research scientists or engineers. At the same time there is great
concern that the parts of our system that have led to the successes of American
science and technology not be destroyed. Indeed, a significant part of the current
debate on calculus reform (a term much disliked by parties on all sides of the
issue) concerns how best to achieve both of these goals. In the current environment, mathematics departments are under pressure to succeed at both. Further,
we must also integrate into our courses the use of powerful computational tools
and somehow do all of this without increasing either course credit or the amount
of time students spend on our classes.

The educational programs in mathematics at large research universities such
as ours, the University of Michigan, divide naturally into three pieces according
to the interests of the majority of the students at that level: the freshman-
sophomore program, the junior-senior-masters program, and the doctoral program. The doctoral program (and the small undergraduate honors program)

receives a lot of faculty attention, primarily because the courses and research experiences given to its students are very close to faculty interests. There is no doubt that this attention is deserved because this high-value program produces the future mathematics researchers and university teachers. Freshman-sophomore programs also receive a lot of attention from departments because of their size and because they "pay the bills." At Michigan, 80% of our departmental teaching load is in the freshman-sophomore courses and 97% of the students in these courses are not mathematics majors. Furthermore, while research is a very important part of faculty work, revenues derived from external sources for research support account for a small percentage of the financial support of mathematics departments; most resources come to us because of our role in teaching mathematics. Of course, the freshman-sophomore programs are also important educationally because of the role of mathematics as an enabling discipline for all of science and engineering, in addition to many other subjects.

Our focus in this article is on the intermediate level, the program for math majors and advanced students from other disciplines. In particular, it seems to us that, especially at large universities, the undergraduate mathematics major who is not heading toward graduate school is the forgotten student. These are the majority of our math majors: at Michigan about 10% go on to graduate school in mathematics and another 30% go to graduate school in another discipline, e.g., law, medicine, industrial engineering, biostatistics, etc.; this leaves 60% who go directly into the workforce. Of these, roughly one-fifth go into K–12 teaching.

In particular, if we are to have workers with high levels of technical skills in the major industrial sectors of our economy, including computing and information technology, banking, insurance, and communications, then many of these positions should be filled by mathematics majors. Indeed, that is where many of our current majors go. We consistently hear from our alumni in business that the level of technical skills needed in many occupations is increasing. As one example, various kinds of contingent securities (or "derivatives") are now commonly and increasingly used in most large businesses. While we may produce enough "rocket scientists" to design and price such financial instruments, do we have enough accountants with sufficient technical background to quantify and independently evaluate their risks? When problems arise, lawyers involved in litigation need a good understanding of the technical aspects of their cases. The mathematics major is a natural arena where students can acquire the needed skills. We should view it as an important part of our mission to encourage more young people to become math majors and to help them obtain in addition to mathematical training the broad liberal arts background that helps them succeed in such careers.

Adopting this point of view raises several difficult and important questions. What exactly are the needed technical skills? What exactly is the mathematics we should teach? Do our current programs do a good job of teaching these skills and topics? In what ways can the special resources of research universities be brought to bear on the problem? How can we communicate to our undergraduates the value of majoring in mathematics?

Except for basic computer skills (programming, use of spreadsheets and word processing software) particular technical skills do not seem to be high on the wish-lists of prospective employers. Rather, the ability areas most often mentioned to us by alumni in the field are problem-solving, breaking complex problems into solvable pieces, adapting a solution from one problem setting to another, writing and otherwise communicating with clients and co-workers, decision making, working well with others in a team, and a willingness to work hard. Many of these are standard components of an undergraduate mathematics curriculum. Much of mathematics deals with solving problems, and usually requires breaking them down into smaller pieces. The abstraction of mathematics is exactly designed to extract principles common to many contexts. Probability and statistics deal with making decisions in the face of uncertainty. Courses in mathematical modeling, which our students almost universally recognize as valuable, deal with choosing the right tool to solve a problem.

Thus it seems that the basic content of our undergraduate courses is appropriate to these new, or at least newly recognized, goals of the undergraduate program. What may be less appropriate or adequate is our pedagogy. Most of us who are mathematics faculty learned throughout our education, but particularly in graduate school, that mathematics is a solitary occupation. We were probably good at timed exams, and when we didn't understand something, we would think and read much more before we would ask for help. These are good approaches for research mathematics, but they are not the way a B.A. mathematician works in industry. In all of the many recent conversations we have had with our alumni in the workforce, we have not found a single example of a work situation that resembles a typical course exam or a problem that has a clean, unique solution in the style of the standard textbook exercise. All of these alumni stress almost as a mantra the importance of teamwork and communication — indeed, these abilities are rated above the raw technical knowledge that we are constantly struggling to impart.

A common element of calculus reform projects has been the introduction of cooperative learning. In the first-year courses at Michigan most submitted homework is done in groups, and much lecturing is replaced by group exercises in the classroom. It is surely time that some of these strategies begin to be adopted also in the upper-level courses. Of course, there will and should remain a good deal of individual work, but we are convinced that we should continue to strengthen students' ability to solve problems cooperatively throughout their undergraduate curriculum. Changes in the first- and second-year courses have also increased the emphasis on communication; gone are the days when a scrawled formula or number counted as a solution to a problem. In most instances in these courses we now require a coherent explanation of the solution in full, good, English sentences. But again, this change has only begun to percolate through the upper-level courses. It is not too much to ask that every course should include at least a small requirement of careful written work.

These changes in undergraduate pedagogy are not easy to make. Faculty of our generation are experienced and comfortable with the lecture model and often at first very uncomfortable with supervising collaborative groups. Teaching communication and writing is difficult and expensive; grading homework and exams with attention to both writing and content is much more time consuming. We will not be able to do it by fiat, but it should be a long-term goal. Furthermore, there are many opportunities beyond the regular classroom to reinforce these lessons. Student research projects, ideally under the auspices of an REU (Research Experiences for Undergraduates) or similar program, are ideal for encouraging both collaborative work and careful written exposition of the results. Although traditional paper-grading jobs are useful for solidifying technical mathematical knowledge, jobs as tutors or classroom aids are far richer in providing experience in communicating mathematics in a cooperative setting.

Our math major program may not be doing a good job of training high-school teachers, judging from what we see of entering university students and what we read in the news. What can research departments do to improve the situation? Probably very little, since it is a large, national problem, beset with local politics and other human issues. At Michigan we have only a small program in this area, and these authors have little experience and no suggestions in this direction. However, the improvement of mathematics instruction and standards in K–12 education is clearly an important problem which should be of concern for mathematicians in research universities.

Another place where we can improve is in the counseling of our majors. The mathematics curriculum is not the only or even the best place to work on developing all of the skills our graduates need; there are many more opportunities for learning than any one department can provide. Mathematics courses take up no more than 30–40% of the junior-senior program for mathematics majors at Michigan. Guidance in the form of suggested programs should be given for many more, perhaps another 40% of their courses. For example, students interested in careers in business, insurance, banking, or information technology should be advised to take courses in economics, computing, writing and speaking, accounting, etc. Students interested in engineering or computing areas should take more courses in computer science, such as databases or operating systems, operations research, modeling, physics, biology, etc. Most students do not have clear ideas of what they want from a program, or which courses to take, and will appreciate having clear recommendations. Furthermore, few employers of bachelors degree students are looking for graduates with specific mathematical skills. For most students, having a broad education that cuts across many disciplines enhances their job prospects. This doesn't mean having a hodge-podge of random courses on the transcript, but taking a broad range of courses that give a strong knowledge base of complementary skills relevant to a general career direction.

In parallel with efforts to improve the curriculum for the undergraduate math major, we need to address the problem of recruitment. How can we show the sophomore deciding on a major that mathematics is a good choice? One obstacle is that mathematics is hard — there is no way our typical major can learn mathematics without serious study. On the other hand, hard work is a characteristic of every scientific field, and there is no reason to believe that students are afraid of hard work. Indeed, a strong work ethic has a lot to do with career success in any field. We should not weaken our programs or relax standards with the idea that this will attract majors. Rather, our students should be made to work hard and to understand that their efforts will be rewarded. We should make every attempt to find out how our students learn best, encourage them in their efforts, and offer them every opportunity to succeed.

Another obstacle to recruiting efforts is that beginning students, as a group, have little idea of what math majors do other than become teachers. We are handicapped by the fact that there are almost no positions below the Ph.D. level with job title *mathematician*, while titles such as *engineer*, *lawyer*, and *economist* are part of everyone's experience. Furthermore, many mathematics faculty don't themselves have a much better answer to the pervasive question "what can I do with a math major"? Fortunately, the national mathematics organizations, most notably the MAA, have produced several excellent brochures and web pages (some are accessible from the Michigan Department's Web page under Student Resources) with examples of the very wide range of opportunities available to the mathematics major. Furthermore, we have found that alumni are willing and often eager to return to their *alma mater* to tell students of the possibilities in their fields.

Faculty need to become better aware of these career opportunities. Once we do, we have an obvious medium for proselytizing: each year we teach calculus to several hundred thousand students. Current texts often treat many interesting applications of mathematics that show its utility in other disciplines. But do the texts and mathematics faculty point out when the occasion arises that these represent possible career paths for mathematics students? For example, one commonly given application of the integral is to compute present or future values of money. Why not take this opportunity to mention careers in actuarial science, banking, and finance? Optimization problems could be linked to careers in communications and transportation.

Throughout all of these efforts to provide a better undergraduate environment for all of our students, we should not lose sight of whom we are serving. Although our students will be the leading citizens of the next generation, most of them will not be the creative wizards who achieve breakthroughs in their fields. They will take the tools that we give them and use them in ways we might not recognize but which are nevertheless crucial to the evolution of society in the information age. We should not expect them all to perform at the level of honors students, but recognize that what will be important for them is as much the overall structure of mathematical thinking as it is the detailed content of our courses.

PETER G. HINMAN
UNIVERSITY OF MICHIGAN
DEPARTMENT OF MATHEMATICS
2072 EAST HALL
525 E UNIVERSITY AVE.
ANN ARBOR, MI 48109-1109
UNITED STATES
  pgh@umich.edu

B. ALAN TAYLOR
UNIVERSITY OF MICHIGAN
DEPARTMENT OF MATHEMATICS
2072 EAST HALL
525 E UNIVERSITY AVE.
ANN ARBOR, MI 48109-1109
UNITED STATES
  taylor@umich.edu

Contemporary Issues in Mathematics Education
MSRI Publications
Volume **36**, 1999

On the Role of Proof in Calculus Courses

THOMAS W. TUCKER

I would like to consider two questions:

- Should students see proofs in a standard calculus course?
- Should students do proofs in a standard calculus course?

I use the word "should" with all its moral overtones because I think that to many this is as much a moral issue as a pedagogical one. On the other hand, I have a hard time distinguishing moral considerations from considerations of taste, where the dictum *de gustibus non est disputandum* applies. Therefore I will for the most part think of "should" as "need" or "must" or even "does it help". I do believe these questions are at the root of some of the debate about calculus reform, but I have heard very little thoughtful discussion of the issue. I must, however, apologize that what I say is woefully uninformed by research in mathematical education, and all I intend is to begin a dialogue.

Before I proceed I would like to differentiate between two pedagogical uses of proof. The first, which I'll call Proof I, is part of a process of formalization and organization. In this setting, the student is presumed to have an effective, reliable understanding of concepts and results; the goal is to develop a formal language with which those results can be proved true. I would consider the proof that the limit of a sum is the sum of the limits as Proof I. The second sort of proof, which I'll call Proof II, is less formal, and is used to answer a question that is in doubt. Students don't often encounter such questions in a calculus course. Here is a very simple example: if the functions f and g are twice differentiable and their graphs are concave up, must the graph of $f + g$ be concave up? The answer is not obvious if one thinks only pictorially and one could well start looking at examples; but of course, once the question is interpreted in terms of the sign of the second derivative, the question easily yields its answer and a proof.

Should Calculus Students See Proofs?

I will list some commonly given reasons why they should, and comment on those reasons. (I am sure there are other reasons, and I am sure my comments are not without bias. Again, I intend only to begin a dialogue.)

1. Proofs help students understand concepts and believe results. This certainly hasn't been my experience, at least with Proof I. Does it help students to see a proof that the limit of the sum is the sum of the limits? To put this in a historical perspective, would it have helped Newton or Euler? Of course, the context I have established for Proof I assumes effective understanding and belief already. However, I don't think that merely seeing Proof I adds much to understanding or belief. On the other hand, I feel Proof II does help, but I have no solid evidence.

2. It is useful in later mathematics courses for students to see proofs in a calculus course. The obvious question here is: What later course? The half-life of a calculus student is one semester, and well under 10% of the students in a standard calculus class will ever major in mathematics. Whether it is useful or not for potential mathematics majors, we better have a good reason for the other 90% of the class. As for the majors themselves, I am sure there is some value in seeing proofs, but I'm not sure how much. Having seen my best students forget in one semester the derivative of the arctangent, I don't have a lot of faith in the vertical recall of students in vertically structured courses. Will they remember the Mean Value Theorem when they get to a junior analysis course? Will the dim memory of having seen the proof help? Was it worth confusing the rest of the class? Now I'm talking about Proof I. Seeing Proof II may help more in later courses, but not nearly so much as doing Proof II.

3. Proofs are part of students' cultural heritage, which they should appreciate the same way they appreciate the theory of relativity or Huckleberry Finn — even if they don't understand it. I am sympathetic to this view, and I have to admit that ten years ago I used to force my class to memorize the ε-δ definition of limit as a form of poetry. I guess that, as I've gotten older, I've become less impressed with the grandeur of this definition. There is better poetry to be remembered. Or there are Kepler's Laws. Maybe I've just found what I feel are better things to do with the little time I have with my students.

4. Proofs are what we mathematicians do, and students should see what we do. I am not so sure we do spend that much time doing Proof I, except in writing textbooks or teaching. Our research is mostly Proof II. Anyway, although I believe what I do is interesting, I'm pretty sure that most of my students don't. They might be expecting proofs from me because that is what a mathematician does, but they are not looking forward to it.

5. Proofs are beautiful. Some of Proof I is beautiful and some isn't. I don't find the limit of the sum proof especially appealing, but I know a breathtaking proof that if $f' > 0$ on an interval then f is increasing (it does not involve the Mean Value Theorem). I am very sympathetic to this motive, just as I am to the cultural one; they may be close to the same thing. The trouble is that although most people are struck instantly by the beauty of a Chardin still-life, there seems to be much less innate appreciation for beauty in thought. An untrained or inattentive eye misses much but it can still see a lot; an untrained or inattentive mind misses everything.

6. Proofs build character. It is hard to argue with this. I used to tell my students the same thing about techniques of integration. Lots of things build character — pain and suffering for example. I always remember the story I heard at West Point that Patton (I think it was) said he had no fear of war after having to do boardwork in front of a calculus class as a cadet. I'm not so sure that says anything I want to hear about the public's attitude towards mathematics, when even war is to be feared less. There are more constructive ways to build character. Moreover, this is akin to viewing calculus as a filter, and ten years later it should not be necessary to repeat the reasons why calculus should be a "pump and not a filter".

Should Calculus Students Do Proofs?

It may seem odd that I have phrased this as a separate question. Scientists and mathematicians generally believe their disciplines can only be learned by doing. Yet mathematicians have backed so far away from asking proofs of students that it is rare to find the verb "prove" in a textbook or exam problem. Rather, the word is "show", "justify", or "explain why". Students know there is a difference. They associate "proof" with something very formal, like writing the answer in latin, and they are always afraid we might mean "prove" when we say "show". We allay their fears; we talk about proof, but we don't put it on the exam. Is material learned if it does not appear on the exam? Does a tree falling in the forest make a sound if there is no one to hear it?

I find it a sad state of affairs that much of the debate about calculus reform has devolved to arguments about whether students in a calculus course should see formal definitions and proofs, with the tacit admission that it is out of the question for students to write proofs of their own. You know you're poor when you fight over crumbs. How did we get to this point, where we dare not use the word "prove" in calculus problems? Many I am sure will cite the quality and attitude of their students, but blaming students only provides solace to the instructor. I think one reason proof by students has been abandoned is that too often it was Proof I we were asking for. Since we knew students have trouble with proofs, we chose proofs that were more mechanical, such as *epsilon-delta* proofs which mostly involve manipulation of inequalities. One might even call this type of proof remedial. The trouble is that students don't react any better to this type of remediation than to other types. They see proofs as pointless exercises in saying things the instructor wants to hear. One of the important lessons of the Treisman programs for students at risk is that students respond better to challenging problems involving thought than to simple problems involving drill and mechanical manipulation. Maybe if we had been asking more Proof II problems, we would not have given up asking students to do proofs.

Reasons for students doing proofs are no different than the reasons given before for seeing proofs, but this time the reasons are more cogent for Proof II. I really do believe that students doing Proof II are gaining understanding. I am sure that doing Proof II helps mathematics majors for later courses; I even feel that non-majors are not wasting their time in doing Proof II, especially science majors. Students doing Proof II are working much more the way mathematicians do. The beauty of proof is much more likely to be appreciated when the question is in doubt and it is up to the student to grapple with the problem, even if unsuccessfully. Doing Proof II, although perhaps less of a discipline than doing Proof I, is a more constructive way to build character; in any case, the only character built by *seeing* proofs is the ability to sit still while someone is talking to you.

The only one of the reasons for teaching proofs that I'm not so sure about as a reason for doing them is the appeal to cultural heritage. It is possible to appreciate music or literature without composing or writing. Perhaps an attentive students can learn to appreciate mathematical culture by seeing proofs, rather than doing them, and Proof I is as good as Proof II. After all, that is exactly what good popularization of mathematics achieves.

So where are we going to find Proof II problems for calculus students to do? Some areas are replete with good Proof II problems that students can do: geometry, number theory, graph theory. Other areas are not so good. I've always thought linear algebra much more oriented toward Proof I, and for that reason it always seemed odd to me to use it as a vehicle to introduce proof to students, especially when the practical applications of linear algebra are so useful to so many. I believe that introductory calculus has plenty of Proof II problems that students can do. Although such problems don't appear on college exams and in texts, the Advanced Placement exam for years has had a final problem which was usually Proof II, frequently involving a functional equation, and those problems have provided food for thought in countless AP classrooms. Many of the longer projects given in calculus reform texts or supplementary materials are Proof II or could be put in that form. Proof II is often more verbal than algebraic, and the calculus reform emphasis on writing and non-algebraic approaches opens up more possibilities for Proof II. In fact, much of the culture shock felt by students in upper level, theoretical courses may be due more to how little they have been asked to write in words a coherent mathematical argument or explanation, than to whether they have seen formal definitions and proofs.

In any case, at this point I think I owe some more examples of Proof II problems than the one I gave earlier. Here are a few to get started.

(i) If the graph of f is increasing and concave down for all x, then the graph of f has a horizontal asymptote. Prove or give a counterexample.

(ii) The student is given the graph of a differentiable function f on the interval $[0, 1]$ such that $f(0) = f(1)$. Clearly at some points the slope is positive and at other points the slope is negative. Prove that the average of the slopes is exactly zero.

(iii) Prove that the function $f(x) = x^2 + \cos(kx)$ has either infinitely many points of inflection or none at all, depending on the value of k (this was on the 1995 AP exam).

(iv) One could define the derivative of f at $x = a$ as the limit of difference quotients of the form $(f(a+h)-f(a-h))/(2h)$, instead of $(f(a+h)-f(a))/h$. For example, graphing calculators use the first quotient rather than the second to estimate the derivative. Does it make a difference in the definition? If the first limit exists, must the second? If the second limit exists, must the first? Prove or give counterexamples.

(v) Prove that the equation $\sin(x) + x = c$ has exactly one solution, no matter what the value of c.

(vi) (from multivariable calculus) Suppose that $f(x, y) = g(x) + h(y)$ and that $g' < 0$ for $x < a$, $g'(a) = 0$, and $g' > 0$ for $x > a$, and that $h' < 0$ for $y < b$, $h'(b) = 0$, and $h' > 0$ for $y > b$. Prove that f has a local minimum at $x = a$, $y = b$.

Some of these problems are pretty hard, but they are not out of the question for a good calculus student. The point is that problems like these should be in a calculus course, and at present they are not.

THOMAS W. TUCKER
COLGATE UNIVERSITY
DEPARTMENT OF MATHEMATICS
13 OAK DRIVE
HAMILTON, NY 13346-1398
UNITED STATES
    ttucker@center.colgate.edu

Case Studies in Mathematics Education

Contemporary Issues in Mathematics Education
MSRI Publications
Volume **36**, 1999

If I Could Talk to the Animals

DOROTHY WALLACE

The title of this essay is a free-floating reference to conversations I have had with my mathematical colleagues over the last few years about the extent to which the classes we teach within our departments ought to inform and connect with the subjects students are studying in other departments. These discussions often end with a complaint such as, "You can't talk to our engineers, they won't listen." or "The biologists just don't care about the math their students take." or "How can you expect an English teacher to have an idea what kinds of math students ought to know?" I'm not sure whether such comments were made with the specific intention of ending an uncomfortable conversation, but they certainly had that effect. I would also be willing to bet that conversations with the engineers, the biologists and the English professors had stopped sometime earlier, if indeed they had ever happened in the first place.

We are at a critical place in the history of mathematics education. Mathematical thinking has insinuated itself into every scientific discipline, including social science. Science and technology in turn have affected every aspect of human society, including art, literature and music. It would be a particularly inappropriate moment for the community of mathematicians to turn inward upon itself for answers to questions concerning the education of students in all disciplines. We must seek these answers jointly with scholars in every area. Only in this way can we construct mathematics courses meaningful to students with a variety of interests and inclinations. Dartmouth was recently awarded a grant from the National Science Foundation under its initiative, "Mathematics and its Applications Throughout the Curriculum". Because of this work, which is expected to result in the creation or reworking of between twenty and thirty courses in our curriculum, some of us have been having conversations with colleagues all over campus about how various students might benefit from the mathematics they take here. As you will see, the answers and suggestions we have been hearing are creative and often surprising. My intention here is to describe some of the conversations I have had with faculty throughout my college; I think you will be surprised by some of the unexpected answers I got and impressed with the willingness of others to seriously consider this issue. I hope it will motivate you to open the way for similar conversations at your institution.

Like many math departments, for years mine had been attempting to have conversations with the physicists and the engineers, whose need for mathematical preparation of their students was very clear cut. General hostility was the usual outcome. The mathematicians felt that the demand for superficial coverage of topics was considerable; as a result, someone would have to invent a pill for students to take, prior to their arrival at college, which mysteriously deposited a repertoire of familiarity with special functions of every type. This seems to be what would satisfy the professors of engineering. The engineers seemed mystified with the mathematicians' seeming unwillingness to teach much mathematics. You would correctly guess that the "calculus reform" effort exacerbated this argument greatly. So, in the last couple of years, we have taken a different tack in our conversations with engineers and physicists. Instead of asking the question, "What topics should we cover?", we have instead asked these faculty to identify for us the specific ways in which their students fall short of expectations in their sophomore classes.

Initially the engineers said, "The students don't really know what a derivative is." We said, "We have tested them in every way possible and they seem to know it when they leave the calculus class." Further discussion yielded two points of agreement. First, the students do not retain material as well as we would like. Second, the students do not recognize concepts in a new context (which in this case was any context outside the calculus class). Once we agreed on these two problems as the focus of our efforts to improve calculus instruction for physical science students, all of these faculty were able to collaborate together to create a joint syllabus in mathematics, physics, and chemistry. The mathematicians argued for covering less material in order to achieve greater student retention. The physicists and engineers argued for an interdisciplinary approach that would improve the students' ability to transfer concepts and skills from one context to another.

Like many math departments, we had never really had a conversation with the biologists. In our cosmology we had pretty much equated biology students with pre-medical students, whose reputation for insincere interest in mathematics was made for them by their predecessors twenty years ago. The pre-med students had a similar reputation with biology professors. So we now make a distinction between the intellectual endeavor of biology and the presumed attitudes of the majority of inhabitants of poor biology's major. It was also surprising to learn of the level of interest among some biology professors, in particular ecologists and population biologists, in the mathematical education of their students. The biologists said, "We would love to have our students take more math after basic calculus, but (i) none of your courses is directly applicable to our needs. Furthermore, (ii) none of the biology students believe that calculus is in any way pertinent to biology. In fact, (iii) they don't even recognize a derivative when they see one in class."

This gave us three problems to solve, none of which is easily addressed. The last two problems can be directly addressed by an interdisciplinary approach to beginning calculus intended specifically for biology students; this would be similar to the solution agreed upon by the physicists and engineers in the last example. But the absence of courses specifically applicable to the needs of biology students requires entirely new courses. And what will drive the mathematical content of these courses? The biologists said, "All we want our students to be able to do is to understand what a phase portrait tells you and to have a good working understanding of what an eigenvalue of a matrix is." They further pointed out that most topics in the freshman courses, such as methods of integration and properties of special functions and vector calculus, were quite irrelevant to the biologist. Here you see directly reflected the needs of both population biology and ecology. In fact, what you are seeing is a change of emphasis in the fields which compose biology itself.

Oddly, even though mathematicians tend to see mathematics as generally useful, it is the mathematics department which must be convinced to serve the needs of the students in this case. Phase portraits and eigenvalues are topics generally reserved for advanced courses which serve primarily math majors. An adjustment of this sort reverses the somewhat proprietary stance we often take toward this material. We will no longer be able to say that the whole purpose of linear algebra lies in its ability to teach future mathematicians how to write proofs. Nor can we say that the purpose of dynamical systems is to generalize the classical approach to ordinary differential equations. The biologists were very clear about this. They said, "We never met a differential equation we could solve analytically in our research."

The biology professors seem to regard exponential growth and the logistic equation as curious anomalies that serve to lead to the more interesting questions. When asked how they approached eigenvalues they (somewhat sheepishly, I thought) said that they defined the eigenvector to be the limiting population distribution in their model. They weren't too happy about that approach either, but it was all they could do, given the background of their students. As a result of these conversations, the math department is now giving serious consideration to an alternative calculus sequence for biology students whose content would be quite different from the sequence an engineer or physicist would take.

Like many math departments, we rarely consider the needs of nonscience students. We have created several courses with no prerequisites; these are to serve those students desiring breadth of education and increased familiarity with mathematics. But I don't recall ever having a departmental discussion of why an art or literature student would want to take a math course, or how such a course could directly speak to their existing interests in art or literature. To me, this is one of the most intellectually engaging questions of its sort, precisely because it is so hard to answer and also because the answer could not possibly come from within the mathematics department alone.

Most of the time, when I ask this question of humanists, the answer I get is sheer nonverbal puzzlement. The question is as foreign to them as it is to us. Once in a while I get a big clue, though. For example, one person pointed out that the earliest science fiction was written during the Renaissance, and was obviously a direct response to some of the science of the time. Another pointed out that there were several modern authors dealing directly with mathematical and scientific issues in their works, and he wished that when he taught those works he could give the students the necessary insights into the mathematical and scientific ideas. Yet another is troubled by how unaware students seem to be of how math and science have informed musical ideas. Putting these discussions together, we can see several strands.

One pervasive theme is the idea that all of culture, whether it be literature, music, or math, is part of a big picture whose parts are all related. That big picture, and the quality of those relationships, is the natural object of study for humanists, especially scholars of literature and history. Such an inquiry is not possible if the contribution of science and mathematics is omitted. Second, there is the suspicion that in some ways the doing of mathematics is more like the doing of art or music than it is like the doing of science. On the individual level, the conception of a mathematical idea is much like the conception of an artistic idea. The commonality of human endeavor comes to the fore here. The sensation of behaving intellectually like a mathematician, even for a very brief period of time, acquires a value far greater than the value of the mathematics invented at that moment. Finally, there is a perceived benefit in the opportunity to interact with modern scientific and mathematical ideas. The faculty I spoke with in no way expressed concern with the ability of their students to compute compound interest, for example. They were much more interested in thinking about very modern issues such as relativity and chaos theory, because they were convinced that these things have a profound effect on how human beings think of their relationships to each other and to the universe.

We see a lot of articles and discussion right now on the subject of "quantitative literacy"; this phrase nearly causes me to break out in hives because of the Orwellian overtones I detect and also because of some of the texts I have seen produced in its name. The whole concept of "quantitative literacy" is being conjured by mathematicians as the answer to the question, "what should the average citizen know?". I hope I have made it clear in this essay that such a question is wrong-headed. The "average citizen" is not what anyone really aspires to be; it is not in the mission statement of my institution, and I would bet it isn't in that of your institution either. (Dartmouth College aspires to produce average citizens. Please send your check for a hundred thousand dollars to _____ .) If you rephrase the key question, however, to read "how can mathematical knowledge serve every person?", then you have certainly created a question that mathematicians by themselves cannot possibly answer. If you then open this conversation to scholars from other corners of your institution, it will soon become clear that there is a huge diversity of human beings who would all benefit from mathematics for completely different reasons.

To increase the intellectual basis in mathematics for the citizenry of the next millennium, drastic action is needed on our part. The first step of that action needs to be serious discussion among people in all disciplines about how math ought to inform, gain from, and interact with other disciplines. Subsequent action must evolve as a response to this dialogue among disciplines. I look forward to a new type of conference involving people from a broad variety of scientific and other disciplines. This conference would engage in serious consideration of the problems raised here, both for the benefit of all of us and as a visible proxy for a process that must necessarily involve us all. Perhaps then the broad sea of mathematics will begin to take shape as a series of pools drawing students of many different disciplines, according to their needs. We must regard the diversity of all comers not as an obstacle to progress, but as what it truly is: our strongest national resource. We must begin to see that one size most emphatically does not fit all. Only then can we build a curriculum that addresses the problem before us in more than a superficial way.

DOROTHY WALLACE
DARTMOUTH COLLEGE
COMPUTER SCIENCE AND MATHEMATICS DEPARTMENT
HANOVER, NH 03755-1890
UNITED STATES
    Dorothy.Wallace@dartmouth.edu

Contemporary Issues in Mathematics Education
MSRI Publications
Volume **36**, 1999

The Research Mathematician as Storyteller

WILLIAM YSLAS VÉLEZ AND JOSEPH C. WATKINS

The Southwest Regional Institute in the Mathematical Sciences (SWRIMS) was funded by the National Science Foundation (NSF) as an effort to integrate research and education. Part of the NSF's vision is that the researcher should be simultaneously involved in research activities and educational initiatives. One of the purposes of SWRIMS was to bring into focus what these educational initiatives might be, and to answer the following questions: How can a researcher use research, especially personal achievements in research, to further the educational goals of the mathematics community? How can this research be used to motivate our children to pursue higher mathematical studies? One of us addressed these questions in [5], which catalogued SWRIMS activities for the academic year 1995-96 at its three funded sites (University of Arizona, Northern Arizona University and Utah State University), and formulated some ideas for integrating research and education. The present article aims to highlight the products of one of these activities and to remind the mathematics research community of its teaching role in our society and its responsibility for transmitting our mathematical knowledge to the community.

It is the role of teacher that we have come to re-think as we carried out our SWRIMS activities. This role is very nicely articulated by Robert A. Williams, Jr., in his foreword to *The Rodrigo Chronicles* [2]. Williams, drawing from his background as a member of the Lumbee tribe, states:

> In the Native American tradition, to assume the role of Storyteller is to take on a very weighty vocation. The shared life of a people as a community is defined by an intricate web of connections: kinship and blood, marriage and friendship, alliance and solidarity. In the Indian way, the Storyteller is the one who bears the heavy responsibility for maintaining all of these connections.
>
> To be a Storyteller, then, is to assume the awesome burden of remembrance for a people, and to perform this paramount role with laughter and tears, joy and sadness, melancholy and passion, as the occasion demands.

Supported by the National Science Foundation grant DMS-941283.

There is an art to being a Storyteller, but there is great skill as well. The good Storytellers, the ones who are most listened to and trusted in the tribe, will always use their imagination to make the story fit the occasion.

The stories that SWRIMS would like to tell are stories of relevance, of applicability, of excitement, of importance. They are the stories of our adventures as we forged new tools to address old problems or found new uses for old tools. These adventures have all of the high drama that we would expect from journeys into unexplored territory, and yet we as a profession have kept such adventures hidden. We have not been effective Storytellers in our tribes, and our children have been kept ignorant of this important aspect of our culture. So much so that many in our community state with pride how ignorant they are of mathematical reasoning. We, the research community, should develop these stories and begin to tell them. If we do not, then who will?

In this article we describe two stories that SWRIMS developed for the high schools. NSF mandated that SWRIMS activities should include undergraduate and graduate students, high school and university faculty. Groups with representatives from each category were formed at the various universities. One of the tasks of these groups was to re-package mathematical research in such a way that its results could be understood by a much larger community. It was hoped that performing this task would provide a valuable learning experience for everybody in the group. The researchers would learn how better to communicate mathematics while the high school faculty would better understand the utility of mathematics. When the groups felt that they had developed ideas that would work in a high school classroom, these ideas were tried out in the classroom by everyone in the group.

Why such broadly based groups? For the undergraduate and graduate students in a group, high school was not so long ago. They understand high school student capabilities and anxieties. Their closeness in age adds to their credibility. As Storytellers, their mainstay is the joke placed to give levity when the circumstances become unduly serious or tense.

High school teachers understand the class management issues that university faculty rarely, if ever, consider. For the most part, teachers operate under conditions that researchers would find intolerable and do so with grace, imparting few of the frustrations created by these conditions. As a Storyteller, the teacher's genre may be the anecdote or the riddle, a short piece well placed in the classroom hour that ties stray facts together and starts heads pondering.

Researchers often spend years focused on a specialty among many possible scholarly subjects. We become researchers only after many years of training, and we understand that our achievements can have importance beyond our specialty. The researcher's story is an epic.

We will now tell our epic with jokes, anecdotes, and riddles, "using our imagination to make the story fit the occasion."

First Story: Exponential Growth

Looking back over the two years of the SWRIMS project, we realize how uncertain our beginnings were. We had no fixed ideas on how to proceed and no models from the mathematical community to guide us. As we met with mathematics and biology teachers from several of the Tucson high schools and became acquainted with them and with their programs, our confidence in our ability to make an impact waxed and waned. Part of our early activities had us retreating to a familiar place, the mathematics lecture. The Core Group settled on attending a series of introductory lectures by Jim Cushing, a mathematician whose research focus is mathematical ecology and population dynamics.

Our major goal for the semester was to go to the high school classroom and teach. We had the good fortune to meet the mathematics teaching staff at Sunnyside High School. Their interest, flexibility, and enthusiasm convinced us that Sunnyside was a good place to start. The classroom team — professors Larry Grove, Bill Vélez, and Joe Watkins, high school teachers Doug Cardell, Paul Dye, and Jeff Uecker, and university students Tyler Bayles, Martin Garcia, and Dianna Peña — began holding planning sessions in conjunction with Jim Cushing's lectures. We settled on four full-day visits to Sunnyside, each visit separated by about a week. Our plan was to show exponential growth and decay in a wide variety of contexts — simpler at first, more subtle as we continued. These experiences are summarized in the Southwest RIMS Class Notes titled *The Exponential Function and the Dynamics of Populations*.

Sunnyside is a barrio school on Tucson's south side, with a predominantly Chicano population. The University of Arizona is six miles to the north. Very few of these students have managed to make this six mile journey.

Day 0: Our First Story Begins. The high school teachers invited the Core Group to Sunnyside to see how high school life proceeds. Our intention was to be anonymous observers trying to get a feel for the place. As we would later learn, personal daily greetings are a tradition in the mathematics building at Sunnyside. The students found us quickly and immediately began describing their activity for the day. After much design and construction, they had completed two large paraboloids and were ready to place them at extreme ends of a hallway crowded with students. Could they communicate over the din using the paraboloids to direct sound? They had had a successful test earlier in the day, and they were eager to repeat the test in the Core Group's presence. The second test was an overwhelming success. This is how mathematics instruction proceeds at Sunnyside: mathematical concepts are turned into physical reality by the students themselves.

The typical classroom at Sunnyside has multiple levels — students enrolled in Pre-algebra, Algebra I, Algebra II, and Pre-calculus can regularly be seen working side-by-side. All students study the same topics, but the mathematical techniques and goals are specialized to be appropriate for each student.

Day 1: Powers of Two. Our first aim was to experience the exponential function — both growth and decay — in a variety of ways. To make the case as simply as possible, we began by suggesting activities that generate powers of 2.

Our first exercise was to successively fold a sheet of paper in half. How many sheets thick is the paper after n folds? Some pre-algebra students, not looking for a formula or pattern, counted the sheets after each fold. The advanced students wrote a simple statement: "After n folds, the number of layers is 2^n." The pre-algebra students ask, quite correctly, "What does it mean to have 1024 sheets after 10 folds when I can't even fold the paper in half eight times?" The ensuing discussion give us our first glimpse into the applicability and the drawbacks of a mathematical model.

One student hustled through the exercise, showed us his result, and then asked us, "What mathematics do I need to study if I want to be an economist?" He had been holding this question for a long time, waiting for just the right chance to ask it. We had brought a university catalog for just such an occasion, and went over the program in mathematics and economics at The University of Arizona, where he is now a sophomore.

The second activity uses a probe for measuring distance and a computer to give a graph of the distance between a moving person and the probe. The graphs were displayed on an overhead projector as they were generated. Because the use of a distance probe is likely to be a new experience for many students, we began with a series of warm-up questions and kinesthetic answers to help them become familiar with this technology. The students were slow to start participating, but soon wanted to create their own graphs. Very soon, every student could describe and create a graph using concepts usually introduced in a calculus class.

When the time came to perform the exponential walks, 2^t and 2^{-t}, the students quickly realized that such walks cannot be sustained for any length of time.

One student, whose friend was having a baby, had heard that a fetus doubles in size every month. Using a birth weight of about 8 pounds, we can fill in a table of monthly weights. Guided by this table, the students embarked on a sophisticated discussion of pregnancy. Everyone had been close to at least one classmate who is now a parent, and so the exponential function had meaning for them.

Day 2: Random Models of Growth and Decay. Our second visit to the high school classroom was devoted to more experiences with the exponential function, focusing on models with some random element, so that we no longer had exact powers. Throughout the classroom hour we conducted two activities simultaneously — the growth of bacterial colonies and coin experiments.

One of the undergraduates, Tyler, was the "biologist" in charge of the experiment. During each of the six class hours, he chose a group of laboratory assistants. We grew two colonies of a locally gathered soil bacteria, one at room temperature, and another at the optimal temperature for this strain. We ran the experiment throughout the day, generating a long table of spectrophotometer readings on the blackboard. For some students, this connection between mathematics and biology was the best part of the whole experience, and many would drop by between classes to examine the data.

At the same time that the bacterial growth experiment was proceeding, the students were also performing two experiments, one on growth and one on decay, in which coins were used to simulate simple branching processes with small but positive variance.

Graphs of the experimental results convinced the students that the growth or decay was exponential. Eventually, several students could make a confident conjecture on the growth rate and give an explanation based on an intuitive understanding of the probabilistic mechanism underlying the experiment.

By this time the Core Group knew, through many personal interactions, that Sunnyside was filled with talented and creative students. Many students did not know how remarkable they were. The Sunnyside teachers wasted no opportunities in arranging some unstructured time for the Core Group members to meet with these students.

One Chicana, a truly remarkable young woman, had just returned from a year abroad in Germany and we conversed for a while in German. She excelled in mathematics, but her love was the theater. Her parents had strongly objected to the year abroad; many families feel that a child who physically leaves the community also leaves spiritually. On the other hand, someone returning with gifts like hers can assume a paramount role in this barrio. This was a story of "kinship and blood". Our role was to reinforce this student's sense of self-worth, so that she could gain the resolve to make her own decisions.

Day 3: Population, Plenty, and Poverty. The goal this day was to develop ideas on the purposes of mathematical modeling and gain some understanding of the limitations of models, using human population dynamics. We handed out copies of [3], which gives short introductions to the lives of six families, from Kenya, China, Hungary, India, Brazil, and the United States. We provided context for each story with a summary on each country, obtained from the *CIA World Factbook*.

Many Sunnyside students live in large families, which are important in their lives. So the variety of models for family planning and population growth fueled an interesting discussion, which in turn led the students naturally to the role of mathematical models. What can these models do for us? First of all, in creating a model, we decide what concepts are important and we state our beliefs on the interactions of these concepts. We can use mathematics to study our model and computers to simulate it. We can test our hunches and make informed conjectures. Models can reveal what is important and suggest what to investigate further. We can use models to assess the impact of competing strategies.

We ended by building a mathematical model. Our population would have two age groups — children and adults. The dynamics would be given through a birth rate, a maturation rate, and a death rate. We then proceeded to simulate the model dynamics.

Some of us were designated children, some became adults, and some were consigned to limbo. There was also a census bureau: a "stork" to compute how many babies to bring from limbo, an "escort" to count how many children should become adults, and a "grim reaper" to compute how many adults should die and go to limbo. The grim reaper was never hard to recruit.

Day 4: Models of Populations.   Our fourth day was devoted to investigating the model we had developed at the conclusion of Day 3. The rates were chosen so that the eigenvalues for this system were 1 and 1/2. This gave us a constant population total and displayed clearly the role of the eigenvalue 1/2 as the rate of convergence to the eigenvector having eigenvalue 1.

We imagined a large spaceship with 40,000 adults and no children headed for Alpha Centauri and looked at the population through 24 censuses, using either a programmable calculator or Stella, a computer software tool. After working with the data, the students could see that powers of 1 and powers of 1/2 had something to do with this model.

The next question turned out to be quite challenging: If we know the initial population, can we predict the stable population? To address this question, each group picked an initial population of 10,000, variously distributed between children and adults. They reported their stable population — 4000 children and 6000 adults in every case.

Next we varied the initial population sizes and recorded the stable populations. The students still saw no pattern. At this point, the Core Group had a caucus. Eigenvectors, we believed, were a natural concept. How could we get this point across? In a near panic, we suggested that they run the model a third time, and record the stable population in a pie chart.

Ahah! All the pie charts looked the same. So, at Sunnyside High School, eigenvectors are understood as stable pie charts. This experience and the ideas are described in [6].

The students then changed a parameter and were able to predict the stable growth rate and the stable pie chart for their new model.

In this hubbub of activity, we end our first story. We began by folding paper and ended by simulating age structured populations. The Sunnyside students were ready for more. An account of the activities they saw as possible continuations could very well be a list of projects drawn by a researcher in population biology.

An Interlude: Exchanging Stories

The midpoint of two year plan of SWRIMS outreach activities on population dynamics was a conference organized by Jim Cushing of the University of Arizona and Jim Powell of Utah State University. One purpose of this conference was entirely typical — to bring together mathematicians who share research interests and have them report on recent findings. However, this gathering had an additional purpose. The presenters were informed that high school biology and mathematics faculty would be attending, and were asked to prepare their remarks with this larger audience in mind.

On the whole, the lecturers should be praised for their efforts and their successes in communicating their ideas. One researcher remarked: "I am inviting high school teachers to all of my conferences — the extra effort made in preparing the talks made them more accessible and enjoyable for all of us."

Many talks could be adapted to provide an exciting experience for high school students. Particularly noteworthy for our story was the talk presented by James Matis of Texas A&M's Statistics Department. He had developed a probabilistic model for the migration of the Africanized honeybee and used it to make a highly successful prediction of the Africanized honeybee's arrival in the United States through south Texas.

Sunnyside teachers Doug Cardell and Jeff Uecker labeled this talk a "must do" and we arranged for them to talk to Matis and begin preparing his ideas for a high school audience.

Second Story: Bee Population Dynamics

Our second story begins in 1956 in Rio de Janeiro, where a bee researcher hoped he could combine the aggressive characteristics of the African honeybee with the nectar collecting predilections of its European cousin, brought to the Americas in the sixteenth century by the Spaniards, to produce a bee that aggressively collects nectar. During this research, 26 colonies of African honey bees escaped.

Soon after the escape, European and African honeybees began to interbreed, and the resulting *Africanized* honeybees took on the behavior of their African ancestors. Over the succeeding decades, the Africanized bee has been spreading into regions where mean daily temperatures remain above $50°F$ most of the time. Their migration front has been moving northward at an average speed of 400 miles per year. By 1990, the time and place of their arrival in the United States of America was the subject of scientific research. James Matis gave us the results of his investigations [4].

Joe Watkins, a mathematician in the University of Arizona Core Group, recruited a team to present these research ideas. Undergraduate Tyler Bayles made a return visit to the SWRIMS project. Robert Lanza, a geography graduate student, was brought on the team to work with the bee biologists to design remote sensing maps for use in predicting migration. Rhonda Fleming represented a fine group of environmental biology teachers at Tucson High School.

The tremendous stroke of fortune was the discovery that Tucson hosts one of the five United States Department of Agriculture's bee research centers. Gloria DeGrandi-Hoffman, an entomologist from the Carl Hayden Bee Research Center, works on models for the dynamics of hive populations and on mechanisms for the Africanization of European honeybees [1]. This "bee team" met weekly at the center, from October 1995 to January 1996, to design the activities.

Our goal was to connect many of the broad themes found in a high school environmental biology course to a common topic — the honeybee. For high schools, the novelty of the activities is that mathematical ideas are used to provide insights into biologically meaningful questions. We hoped to intrigue students with current research questions in the biology of bees. The results of the six days of activities are documented in [8]. We wanted to assess our activities with a broad range of students. Thus each day's activities were tested in six different classes at Tucson High School — one regular class, two with at-risk students, two with predominantly Spanish speaking students, and one with many high achieving students — with "bee team" members presenting the activities. We made our visits on consecutive Fridays.

Each day's activities began with a reading. Team members composed these readings by gathering information from beekeepers' handbooks and circulating it around the bee lab for updates and clarifications.

During the class, we discussed the biology of bees extensively, and presented the mathematical activity in clear and concise language. Students formed their own working groups to formulate pertinent questions and to seek out methods of investigation. This method freed the teachers to move among each group of students and to work with them as they addressed the issues and prepared their presentations for the end of the classroom hour. These presentations revealed that many imaginative methods are possible.

These Fridays could be quite intense and the teachers (Oscar Romero, Richard Govern, Kay Wild, and Rhonda Fleming) made careful preparations before and extensive follow-ups after so the day would go well.

Day 1: Sizing up the Population. Working methods for obtaining a reliable census differ dramatically depending on the animal or plant whose population is being tracked. For bees, this can be accomplished by photographing the frames in a hive and estimating the number of bees or brood in the photograph. This task in estimation was the first day's activity. We kept the mathematics gentle because we wanted to develop the idea that mathematics and biology are two subjects that can work side by side.

Day 2: Practical Knowledge of European and Africanized Honey Bees. In the college classroom, the instructor can stick to a syllabus. In high school, the students' demand for a response to all of their concerns is much more pronounced. Some students could not study mathematical models until their doubts about Africanized honeybees were addressed and they were told how to deal with possible stinging incidents. Using materials prepared by the Carl Hayden Bee Research Center, we presented practical training to several hundred students at Tucson High School. In this way, we could advertise our presence and give a context for discussion of bees outside of class.

We addressed the practical side of human co-existence with the honeybee from the viewpoint of the gardener, the pet owner, the outdoor adventurer, the consumer of food, the homeowner, and the beekeeper. Now that a significant fraction of the southern United States is populated by the Africanized honeybee, we must add some scientific knowledge to our practical information.

This training had more relevance than anticipated. One student realized during the presentation that she had encountered a swarm of Africanized bees just to the west of campus. They were safely removed later that day.

Day 3: Hands on Models of Population Dynamics.   The exponential and the logistic curves are ubiquitous in describing population growth. On this day, we tried to explain their widespread appearance by describing the elegant mechanisms behind them. We engaged in two activities using pencil, paper, styrofoam cups with lids, and black and white beads to generate exponential and logistic curves.

On this day, we also had to deal with the tragedy of a drive-by shooting. In this case, the victim had nothing to do with the altercation. Students quite naturally had priorities other than playing with beads, recording data in tables, and making graphs. Today a part of our story had tears of sadness and the connections of friendships. The storyteller's skill is needed in respecting this saga of violence and tragedy. We have seen the high school teacher use this skill far too frequently.

Day 4: A Month in the Hive.   We asked a lot from the students for this day. Before our arrival, they took their knowledge of bee biology and consolidated it into a flow diagram. Most of the day was dedicated to navigating through the technological apparatus necessary to see how our model predicts the change in a hive population over a month's time.

The effort had its reward. Each pair of students chose a month in the year, a city in the United States, an initial population of brood, house bees, and foragers, and a reasonable guess on the abundance of nectar and pollen. By magic, the three equations used for the model produced a biologically meaningful result over and over again.

This was an entirely new experience for them — mathematical biology had something to it. Concepts from biology could be turned into equations and equations can be used to make predictions. These equations were our best shot at identifying the important concepts. Constructing the model forced us to clarify the relationship between concepts and to choose the most important relationships. The mathematics itself was telling the story.

Day 5: A Year in the Hive.   On this day we used a simplified version of BEEPOP, a model produced by the Carl Hayden Bee Research Center, to model a year in the life of a hive. We called our model BEEPOPITA.

This exercise helped students appreciate the effort of ecologists, meteorologists, economists, and epidemiologists in designing and evaluating a model. In addition, we hoped that these students would now be able to evaluate the quality of the predictions they encounter in the popular media. Many models are not based on good science or do not use appropriate mathematical methodology. Many journalists do not have the necessary expertise to make a reliable report of scientific research.

Finding the appropriate balance between mathematics and biology is the continuing challenge in constructing a mathematical model. If the students set out with a model that includes *all* of the concepts in their flow charts, they will be quickly overwhelmed by computation. If their reaction has them removing too many concepts from the biology, the mathematics will be easy. However, the information will not say much about bees (populations, the weather, the economy, or an epidemic).

Day 6: Birth, Death, and Migration.   BEEPOP is a model that works well when resources are available, and the hive does not face any circumstances that are overly stressful. These conditions cannot continue forever. At some point, the hive will become overcrowded or will be disturbed. Environmental conditions may change — the hive may be under attack by a bear, by ants, by a pesticide, by disease, or by fire. The queen may no longer be productive. In these instances, the colony must make a critical decision.

The colony of bees can divide or abscond. It can supersede an unproductive queen. If it does nothing, the hive will die.

James Matis, the statistician from Texas A&M, turned the rates at which these critical events happen for Africanized honeybees into a model based on local environmental conditions. Robert Lanza, the geography graduate student on the "bee team", incorporated these model parameters to design a false color composite map of Arizona.

Because the colors are not what the eyes see, some experience with false color composite maps is necessary before they reveal their relevant information. We began our classroom activity by projecting a slide of a false color composite map of the central part of Tucson. After a while, students could obtain information from the map and find significant landmarks in their lives — parks, malls, schools, and their own homes.

Bees and humans differ in their views on which landmarks are important, and so the remote sensing map of Arizona requires a bit more examination before it is useful. However, both bees and humans agree that the rivers, the mountains, the deserts, and the Grand Canyon are important landmarks in Arizona. We also presented a false color composite map of the Mexican state of Sonora. Many Tucsonans have connections to Sonora and know its geography. Also, the Africanized honeybee made its arrival to Arizona via Sonora, and so we can learn about the possibilities of migration in Arizona from the information collected in Sonora.

The birth-death-migration model developed by James Matis and Thomas Kiffe in [4] used the types of models that we had previously seen using beads and styrofoam cups. We added to this model the process of migration — modeled by a random walk. For Matis and Kiffe, this is a two dimensional process. In Arizona, because the movement of the Africanized honey is along riparian areas, the process is one dimensional, and hence easier to simulate.

What are the parameters in our simulation? We can learn this from migration data along the rivers in Sonora. Similar regions in Sonora and Arizona will be revealed by the remote sensing maps. Could we obtain these data? To date, the answer is "no". We know all the ingredients to make the model, but we can not predict when the Africanized honeybee will first reach Hoover Dam until we put our hands on these data.

A Final Note.   The bee team met with a purpose: to create a valuable educational experience for high school students. As a consequence, however, the team will also have a significant research achievement.

Gloria had been working on queen development time as an explanation for the Africanization of the honeybee population. By July, 1996, the data were in, but the statistical analysis was not revealing the secrets that the researchers knew were true. In a one hour meeting, one of us (Watkins) and DeGrandi-Hoffman were able to work out the model that established this fact — any place that the Africanized honeybee can live, it will take over. Invasion of species and the inheritance of complex characteristics are basic questions in ecology. This question can now be informed by the example of the Africanization of the honeybee population and the mathematical model that comes with this example.

All of us have faced the embarrassing moment after new acquaintances learn that they are conversing with a mathematician. They feel no connection to us — mathematicians are a distant and foreign breed. Now, when we meet someone, we have stories to tell of friendship, alliance, and solidarity. The response to these stories is always warm. We plan to have more stories to tell and we encourage you to find yours.

Years have passed since we began this activity. Students, beginning with the very first group at Sunnyside, still run to meet us and to tell us how much they enjoyed our visits. By telling stories of mathematics, we set out to change the way people view the world. People's views did change, beginning, most pointedly, with our own.

References

[1] G. DeGrandi-Hoffman, S. A. Roth, G. Loper, and E. Erickson, "BEEPOP: a honeybee population dynamics simulation model", *Ecological Modeling*, **45** (1989), 133–150.

[2] Richard Delgado, *The Rodrigo Chronicles: conversations about America and race*, New York University Press, New York and London, 1995.

[3] Paul Ehrlich and Anne Ehrlich, "Population, plenty, and poverty", *National Geographic Magazine*, December 1998.

[4] J. H. Matis, T. R. Kiffe, and G. W. Otis, "Use of birth-death-migration processes for describing the spread of insect populations", *Environ. Entomol.* **23**.

[5] William Yslas Vélez, "The integration of research and education", *Notices Amer. Math. Soc.* **43**:10 (October 1996), 1142–1146.

[6] William Yslas Vélez and Joseph C. Watkins, "The teaching of eigenvalues and eigenvectors: a different approach" preprint.

[7] Joseph C. Watkins, "The exponential function and the dynamics of populations", class notes, Southwest Regional Institute in the Mathematical Sciences, University of Arizona.

[8] Joseph C. Watkins et al., "BEEPOP: The dynamics of the honeybee populations in the hive and in the wild", class notes, Southwest Regional Institute in the Mathematical Sciences, University of Arizona.

WILLIAM YSLAS VÉLEZ
UNIVERSITY OF ARIZONA
DEPARTMENT OF MATHEMATICS
TUCSON, AZ 85721-0001
UNITED STATES
  velez@math.arizona.edu

JOSEPH C. WATKINS
UNIVERSITY OF ARIZONA
DEPARTMENT OF MATHEMATICS
TUCSON, AZ 85721-0001
UNITED STATES
  jwatkins@math.arizona.edu

Contemporary Issues in Mathematics Education
MSRI Publications
Volume **36**, 1999

Redesigning the Calculus Sequence at a Research University

HARVEY B. KEYNES, ANDREA OLSON, DOUGLAS SHAW, AND FREDERICK J. WICKLIN

ABSTRACT. The School of Mathematics at the University of Minnesota is developing a new calculus sequence for students in mathematics, science, and engineering. The sequence incorporates changes in content and methods of instruction. The students are routinely asked to work together cooperatively in small groups. An innovative feature involves student exploration of mathematical ideas and complex, open-ended, interdisciplinary applications using interactive features of the World Wide Web. This paper describes some of the special features of the sequence, including some different approaches to instructional teamwork and student-centered instruction. Student attitudes about the usefulness of the pedagogical and curricular components, and how these approaches affect their learning, are analyzed for this new sequence. Quantitative data are presented that compare the achievement and retention of students in the new sequence with a control group from the standard calculus sequence. Future directions for refinements in teaching the sequence, new curricular approaches, and additional statistical information are discussed.

Introduction

The major goal of the University of Minnesota Calculus Initiative is to create a challenging sequence for mathematics, science, and engineering students in which students obtain a better understanding of how to use calculus as a tool both for mathematical analysis and for solving problems in other disciplines. The new sequence is primarily intended for the middle 67% of the calculus-ready science and engineering students, and for some social science majors. The Initiative is a major reconceptualization of the traditional large lecture/recitation approach used in many doctoral institutions, and incorporates key aspects of successful calculus reform efforts in a new instructional format and supportive learning environment. This reorganization includes changes in curriculum, content, and methods of instruction supported by the use of technologies, and in pedagogy that encourages student-centered learning by increasing student/faculty interaction.

Interdisciplinary or applications-based projects and labs are a central component of the Initiative. They engage students in the analysis of science or engineering-based problems that interlace mathematical concepts with important scientific and engineering applications. Student teams investigate these applications in a workshop setting. The teams collect and analyze data, formulate and test conjectures, and communicate their ideas through oral and written reports. These investigations require students to become more actively involved in the learning process and help them to meet the increased responsibility for acquiring routine computational skills. Improvements in student attitudes about mathematics and the use and value of mathematical reasoning in other disciplines are anticipated outcomes. The materials have been designed so that strongly motivated students can remain challenged and able to explore topics in great depth, while less well prepared students can focus on the mastery of essential concepts. All layers of the materials foster teamwork and stress the importance of clear communication of mathematical ideas. Ideas for the applications-based projects were developed through discussions with science and engineering faculty. The content and technical aspects of the projects were created as a cooperative effort with the Geometry Center. The technologies used vary as the sequence progresses. In the first year, students extensively use graphing calculators with a few forays into labs that utilize interactive hypertext programs that run over the Internet. During the second year, students use Maple and Matlab in regular laboratory settings, although the essential components of these materials can be made accessible to classrooms whose technology is limited to graphing calculators.

In principle, the new calculus sequence can be implemented with minimal resources beyond the traditional large lecture model (explained below). The current model has 100 students per professor, but the students meet in small classes of 25 during 3 of the 5 hours each week. The workshop setting has created excellent opportunities for extensive small-group activities and experiences, and provides the overall atmosphere of a small classroom. The additional instructional cost has been relatively modest, approximately $225 per year per student.

This paper provides a careful analysis of the quantitative and qualitative student outcomes that the new sequence hopes to achieve. Outcomes are compared with a control group of 240 mathematics, science, and engineering students with similar placement scores who participated in the standard calculus sequence. Quantitative data about retention, achievement, and successful completion of the entire first-year calculus sequence is developed for both groups. Qualitative data about student attitudes regarding the usefulness of the structure of the sequence, pedagogical approaches, and the content of the sequence are examined for the Initiative students. This includes initial positive results about homework and group work as enhancements to learning.

Implementation

The curricular and structural aspects of the Initiative are described in this section.

Curriculum and Expectations. The primary goal is to have the students leave calculus with the mathematical skills they need to succeed in future science and engineering courses, and ultimately in their careers. Secondary goals are (1) to motivate students to remain in the calculus sequence and in the Institute of Technology (IT), (2) to encourage the students to become more responsible for their learning, and (3) to help the students understand the usefulness of calculus. The six distinct features of the curriculum that address these goals are described below:

1. *Mandatory Computational Proficiency.* Students may not receive a course grade until they have successfully passed a gateway exam, computing 8 of 10 standard derivatives (first quarter) and integrals (second quarter).
2. *Type of Text.* The text is A. Ostebee and P. Zorn, "Calculus from Graphical Numerical and Symbolic Points of View," a reform text emphasizing geometric reasoning, visualization, and conceptual approaches and written at a level that students can read and understand before going to lecture.
3. *Emphasis on Concepts, Geometric Reasoning, and Visualization.* Students are required to go beyond algebraic manipulations, and use geometric and visual reasoning to analyze complex functions and functions defined solely by graphs. This expectation of conceptual understanding is reflected in the examinations, the homework, and the projects.
4. *Communication of Mathematical Ideas.* Students are expected to present mathematics professionally, both orally and in writing. They give solutions in class, write up certain homework problems using complete sentences, clear exposition, graphs, and diagrams to support the solution, and develop project reports each quarter.
5. *Integration of Applications.* Real-life problems and examples are central to the course, in contrast to the traditional approach of giving special "applications problems." Application modules include topics from: optics (mathematics and the geometry of rainbows), limnology (analysis of CO_2 production in fresh water systems), civil and mechanical engineering (static deformation of cantilevered and simply-supported beams), and biology and ecology (population dynamics using dynamical systems).
6. *Problem-Solving.* While template and drill problems are used to develop skills, the majority of the problems for students require a synthesis of the ideas con-

tained in the course. The homework, tests, and quizzes emphasize synthesis problems.

Structure. The four main components of the new course are the lecture, the workshops, the lab, and the homework. These have been designed to reinforce each other, and to encourage and support active learning.

Format of Lecture. Twice a week, the students attend a one hour large class session conducted by the professor. The main purpose of these sessions is to present new concepts, to provide mathematical explanations of the concepts introduced in the workshop the previous day, and to provide guided practice for homework problems.

Format of Workshops. The students attend one 50-minute workshop and one 100-minute workshop per week; these workshops have been designed by a lead workshop instructor to ensure continuity across the sections. The course professor or workshop leader assists the instructor during the 100 minute workshop. One workshop features problems at the average level of the homework problems, allowing the students time for guided practice. Activities in the other workshop foreshadow future concepts, allowing the students the opportunity to develop computational expertise for upcoming theory. Important aspects of both types of activities are effective communication and the development of problem-solving abilities with peers.

Format of Labs. Four times during the first year, each student works—in a team of three or four members—through large scale applications using the mathematics he/she has learned. Although the students spend only a few class sessions working with their teams, the labs are available over the World Wide Web, allowing the students to continue out-of-class work from their home, dorm room, or a campus terminal. Initially, hard copies are distributed in class; these are followed by a workshop that meets in a computer lab. In the lab, students run the experiments and simulations and get assistance as needed. Each team develops a lab report over the next few weeks outside of class in which ten to twenty extended questions are addressed. In addition to learning some non-trivial mathematics, the students practice being part of a research team: allocating efforts, overseeing quality, making design decisions, and communicating final results.

Findings

To substantiate which student outcomes were influenced by the pilot sequence curriculum and pedagogy, a control group of 240 students was created from the IT students enrolled in the standard calculus sequence. For sampling consistency, the control group students were selected primarily on the basis of their placement exam scores and secondarily on their status as incoming Institute of Technology freshmen. A database was developed to track students in the control group throughout the standard calculus sequence and their upper division engineering and science course work. A parallel database has also been created for the pilot sequence students. While the overall placement score pattern is similar, the cohort distributions do indicate some differences. Thus, direct comparisons may become more reliable when looking at sub-cohorts (e.g., all students with a particular placement exam score) than across the entire population.

Achievement. The percentage of students in the pilot group who successfully completed (achieved a final grade of "C" or higher) each quarter of the first year of the sequence averaged 97% compared to the control group average of 85%. Especially significant was the grading process and the number of high grades ("A"s). The pilot sequence set higher expectations for work, conceptual understanding, and technological proficiency for its students than did the standard sequence, and at least initially demanded much more time and effort. In the first quarter, student adjustments to this increased set of expectations and some unrealistic homework expectations led to a slightly elevated grading process, noting that certain aspects of the student performance (attendance and class participation) exceeded initial expectations and had a positive influence on the grades. In the second and third quarters, as students and faculty expectations stabilized, grading became more standardized, and grades more carefully reflected high expectations and achievement. Thus it is significant that the percentage of "A"s in the pilot courses remained constant at 42-43% throughout the first-year sequence. This average high success rate of 43% is significantly higher than the control group average of 22%. Assigning grade values of $A = 4$, $B = 3$, and $C = 2$, the average grade point average (GPA) for students in the pilot course was 3.31, a difference of .43 compared to an average of 2.88 in the control group, or nearly half a grade level higher in average achievement.

For overall completion, 66% of the students initially in the pilot course successfully completed the pilot sequence versus 62% of the control group students in the standard sequence. More significant was the percentage of students (83% of 97) who successfully completed the Fall quarter of the pilot sequence and then successfully completed a year of calculus (in either the pilot or the standard sequence). In comparison, of the 240 control group students, only 62% successfully completed the first year of calculus.

Retention. Eighty percent of the students who initially enrolled in the pilot sequence have registered for a second year of calculus either in the pilot or in the standard sequence compared to 61% of the control group students. An interesting trend is the consistently increasing quarterly retention rate of the pilot group students—80% of the students who started in the pilot sequence enrolled in the second quarter pilot course, 87% of the students enrolled in the second quarter pilot course took the third quarter course, and 93% of the students enrolled in the third quarter of the pilot course registered for the second year of the pilot sequence; in contrast, retention of the control group students between quarters remained constant at 85%.

Student Expectations and Reactions. Student attitudes about the usefulness of the pedagogical and curricular components and how these approaches affected their learning were collected for the pilot sequence. Student ratings and comments on group work (peer collaboration) and homework were used to determine whether the students became more responsible for their learning, and student ratings and comments on the workshops and the lab projects were used to analyze students' understanding of the usefulness of calculus to science and engineering applications.

Increased Student Responsibility for Learning. An average of 88% of the students rated group work and peer collaboration as a useful-to-extremely-useful element of a calculus course, and 80% rated homework as a useful-to-extremely-useful course component on the survey given during orientation (one day prior to the start of the Fall 95 first quarter). Nearly a quarter of all of the students (24%) stated that group work or collaboration with fellow students was one of the most important elements of a calculus course, and 22% stated that homework assignments were one of the most important elements of a calculus course.

Group Work. On the first quarter mid-term survey, small-group work was rated as useful-to-extremely-useful by only 61% of the students. This increased to 64% on the second quarter mid-term survey and to 78% by the third-quarter mid-term survey, possibly reflecting the growing ability of students to productively collaborate in group settings. Collaboration with peers remained consistently rated at 76%, 74%, and 76% on the mid-term assessments. Students who agreed or strongly agreed with the statement: "Time spent working in groups was worthwhile" formed 60% of the total in the first quarter, 83% in the second quarter, and 89% in the third quarter; the corresponding statistics for the statement: "Working in small groups led to understanding of the subject matter" were 87% in the first quarter, stabilizing to 82-83% in the second and third quarters.

Homework. In the first quarter, the expectations for the homework assignments, including extensive professionally-written problem sets, approached honors-level and competed for study time with other challenging pre-engineering and science courses. Hence only 47% of the students rated the homework as useful-to-extremely-useful on the first quarter mid-term survey, down from 80% on the orientation survey. Adjustments by the instructional team to the assignment load resulted in improved homework quality and student attitudes in subsequent quarters. We noted that 83% of the students rated homework as useful to extremely useful on the second quarter mid-term survey and 76% of the students rated homework as useful-to-extremely-useful on the third quarter mid-term survey. Students who agreed or strongly agreed with: "The amount of homework was appropriate for the course" went from 48% to 81% at the end of the second quarter, and 91% at the end of the third quarter, while the corresponding ratings for "The grading of the homework reflected my understanding" went from 59% to an average of 80% by the end of the second and third quarters.

Increased Student Understanding of the Usefulness of Calculus. Student survey ratings and comments on the workshops and the lab projects were used to quantify the relationship between the students' understanding of the usefulness of calculus to science and engineering applications and the Initiative's curriculum and pedagogy.

Workshops. Students who agreed or strongly agreed with the statement: "The workshop classroom atmosphere was a positive learning environment" formed 88% of the total at the end of the first quarter, 90% at the end of the second quarter, and 95% at the end of the third quarter. The students also gave a high rating to their interaction with the workshop instructors: 94%–100% of the students agreed or strongly agreed that the availability, attitude, and support of the workshop instructors had a positive influence on their learning.

Lab Projects. The survey results suggest that the students believe that the extended projects were worthwhile and worked well for learning both mathematics and real-life applications; some, however, expressed concerns about the extra time required to complete these projects. Overall, 65% rated the lab project as a moderately-useful-to-useful course component. When asked if the lab project enhanced their understanding of calculus and its application, 38% of the students answered "yes", 24% answered "maybe", and 38% answered "no".

Although the students ranked the lab projects as useful overall throughout the sequence, specific items about the labs indicated some differences. While 47% agreed-to-strongly-agreed that time spent on the Numerical Integration Lab was worthwhile, only 36% did so for the Beam Lab. One possible explanation for the popularity of the Numerical Integration Lab was its use in place of a chapter from Ostebee and Zorn, making the reason for doing this project clearer to the students.

The Future

Considering the success of the Initiative's pilot year, two large class sections of the first year sequence will be taught in 1996–97. One of the workshop subsections will be taught in a residential hall to IT freshmen who are part of the University's Residential College program. In addition, some selected liberal arts students with similar mathematics competence and interest as the IT students will be invited to participate.

Each section's instructional team consist of one faculty member, four graduate student workshop instructors, and a fifth graduate assistant in the new role of workshop leader. The faculty member will conduct the two large class sessions per week and rotate between two of the four workshop sessions. The workshop leaders will develop workshop materials and exams, and rotate between the other two workshops. The workshop instructors will conduct two workshops per week and assist in the large class sessions. The instructional teams will meet weekly to ensure consistency across all sections.

The second year sequence includes about 15% transfer students from first-year calculus courses similar to the pilot course, and strong students from the standard Minnesota calculus sequence in addition to students who completed the 1995–96 pilot sequence. Mathematics departments at local state and community colleges have been made aware of this opportunity.

The content of the new second year sequence is multivariable calculus and differential equations using elementary linear algebra, emphasizing geometric solutions of differential equations and building on the conceptual framework established during the first year. Concepts from linear algebra will be introduced as needed for the analysis of curves, surfaces, and systems of differential equations. Students will begin to use higher-level technological tools such as symbolic software and numerical software for linear algebra and dynamical systems. The students will engage in shorter but more frequent labs—4 or 5 per quarter—to explore applications which help students to visualize abstract ideas that arise in the geometry of curves and surfaces. By the end of the sequence, it is expected that the students will be comfortable using symbolic, visual, and numerical techniques to attack complex problems, and to understand when such techniques break down. The five quarter core sequence will prepare students to pursue several options for a sixth quarter, including topics such as complex analysis, vector field theory, or an introduction to numerical methods.

Conclusions

The new calculus sequence for science and engineering students differs from the traditional sequence in content, method of instruction, and in the relative emphasis on group-work, interdisciplinary applications, and technology. The initial achievement and retention findings indicate the following:

(i) 83% of the pilot group students successfully completed the first year of calculus compared to 62% of the control group students,

(ii) the average GPA of successful completion for the pilot sequence students was 3.31 compared to 2.88 for the control group, and

(iii) 80% of the pilot group students are enrolled in a second year of calculus compared to 61% of the control group students.

Pilot group student reactions indicated the following:

(i) The usefulness of small group work (61% rated it as a useful-to-extremely-useful component mid-Fall 95, 64%, mid-Winter 96, and 78%, mid-Spring 96) and homework (47%, Fall 95, 83%, Winter 96, and 76%, Spring 96) increased throughout the quarter,

(ii) The workshops were a positive learning environment (an average of 91% agreed to strongly agreed throughout the year); and

(iii) The lab projects were a useful course component overall (an average of 69% consistently rated them as useful during the year).

The Initiative's new approach to calculus has integrated both content and instructional changes that emphasize active learning. This preliminary evaluation suggests that these approaches are changing student outcomes in calculus at Minnesota and, more importantly, influencing student understanding of how to use calculus as a tool for problem-solving across other disciplines.

Note: Further details are provided in a forthcoming report, which is available from the authors. Additional information can also be found in [1].

References

[1] J. Leitzel, editors, *Assessing calculus reform efforts*, MAA, 1995.

[2] H. Keynes and A. Olson, "Calculus reform as a lever for changing curriculum and instruction", pp. 248–251 in *Proceedings Fourth World Conference on Engineering Education*, St. Paul, MN, 1995.

HARVEY B. KEYNES
UNIVERSITY OF MINNESOTA
SCHOOL OF MATHEMATICS
127 VINCENT HALL
206 CHURCH ST. SE
MINNEAPOLIS, MN 55455-0488
UNITED STATES
keynes@math.umn.edu

Contemporary Issues in Mathematics Education
MSRI Publications
Volume **36**, 1999

Is the Mathematics We Do
the Mathematics We Teach?

JERRY UHL AND WILLIAM DAVIS

In a recent article [7], William Thurston called attention to the sad state of the mathematics classroom:

> We go through the motions of saying for the record what the students "ought" to learn while students grapple with the more fundamental issues of learning our language and guessing at our mental models. Books compensate by giving samples of how to solve every type of homework problem. Professors compensate by giving homework and tests that are much easier than the material "covered" in the course, and then grading the homework and tests on a scale that requires little understanding. We assume the problem is with students rather than communication: that the students either don't have what it takes, or else just don't care. Outsiders are amazed at this phenomenon, but within the mathematical community, we dismiss it with shrugs.

This brings up the question: Does what is taught in the typical mathematics course even qualify as mathematics?

In another article in the same issue of the *Bulletin* [4], Saunders Mac Lane offered:

$$\text{intuition} - \text{trial} - \text{error} - \text{speculation} - \text{conjecture} - \text{proof}$$

as a sequence for understanding of mathematics. In contrast, the sequence in place in most modern mathematics courses is:

$$\text{lecture} - \text{memorization} - \text{test.}$$

Most working mathematicians agree with MacLane's description, thus leaving the inescapable conclusion that the mathematics we do is not the same as what is commonly offered in the classroom. The questions for the next century are:

- Can the mathematics we offer in the classroom be more like the mathematics we do?

- Can we ignite students' mathematical interest?
- What role can computers play in dealing with these questions?

Phillip J. Davis [2] indicates how the answer to the computer question sets up answers to the others:

> The capabilities of all mathematicians are elevated by their association with computation. The transformation by the computer of triangle geometry and of many other areas has, paradoxically, reconfirmed and strengthened the the vital role of humans in the wonderful activity known as mathematics. Put it even more strongly: mathematics develops in such a way that the role of the mathematician is always manifest...
>
> In connection with visual output, I have even argued for the recognition of "visual theorems"... where what the eye "sees" need not even be verbalized let alone formalized in traditional formal mathematical language... subtle feeling that that language cannot even name, let alone set forth...
>
> As regards mathematical education, I think the message is clear. Classical proof must move over and share the educational stage and time with other means of arriving at mathematical evidence and knowledge. Mathematical textbooks must modify the often deadening rigidity of the Euclidean model of exposition.

Calculus&Mathematica as a Prototype Reaction to the Issues and Questions

Calculus&Mathematica [1] is a new computer laboratory calculus course developed at University of Illinois at Urbana-Champaign and the Ohio State University expressly to deal with the questions and the issues raised above. The course is freshly built from the ground up. The purpose of the course, the ways of implanting mathematical ideas into students minds, the ways of motivating students to learn, and the ways of making students retain the important ideas have all been rethought.

As a result, Calculus&Mathematica is the most thoroughly new calculus course available today, and it presents a new model for successful learning of calculationally heavy sciences. Not screened from the essence of calculus by labor-intensive calculations and plots, students in Calculus&Mathematica get right to the good stuff. From the very beginning, they see calculus emerge as the first course in scientific measurement, calculation, and modeling. Students also see calculus as a highly visual and often experimental scientific endeavor just as research mathematics is. The medium is a live electronic interactive text composed of lessons written in Mathematica Notebooks. Each interactive Calculus&Mathematica lesson consists of the following set of Mathematica Notebooks:

- Basics Notebook, for the fundamental ideas,
- Tutorials Notebook, for sample uses of the basic ideas,

- Give It a Try Notebook, for actual student work, and a
- Literacy Sheet: for what a student should be able to handle away from the machine.

For an annotated example, see http://www-cm.math.uiuc.edu/work/examples.

The National Research Council report *Moving Beyond Myths* [5] describes Calculus&Mathematica as follows:

> An innovative calculus course... [which] uses the full symbolic, numeric, graphic, and text capabilities of a powerful computer algebra system. Significantly, there is no textbook for this course — only a sequence of electronic notebooks.
>
> Each notebook begins with basic problems introducing the new ideas, followed by tutorial problems on techniques and applications. Both problem sets have "electronically active" solutions to support student learning. The notebook closes with a section called "Give-it-a-try," where no solutions are given. Students use both the built-in word processor and the graphic and calculating software to build their own notebooks to solve these problems, which are submitted electronically for comments and grading.
>
> Notebooks have the versatility to allow re-working of examples with different numbers and functions, to provide for the insertion of commentary to explain concepts, to incorporate graphs, and plots as desired by students, and to launch routines that extend the complexity of the problem. The instructional focus is on the computer laboratory and the electronic notebook, with less than one hour per week spent in the classroom. Students spend more time than in a traditional course and arrive at a better understanding, since they have the freedom to investigate, rethink, redo and adapt. Moreover, creating course notebooks strengthens students' sense of accomplishment.

Unlike point-and-click multimedia and programs that merely turn pages, Calculus&Mathematica presents examples that can be modified by the student and rerun; so that each example in Calculus&Mathematica is as many active interactive examples as the student wants.

The premise behind Calculus&Mathematica is that if students have the opportunity to go about their work in a manner similar to the manner that working research mathematicians go about their work, they have a good chance for success. Here are some of the principles on which Calculus&Mathematica is based:

Communicate new ideas visually and experimentally; get an idea across before putting language on. Unique to Calculus&Mathematica is the attempt to get mathematical ideas into the students' minds visually before words are put on. To paraphrase Stephen Jay Gould: Scholars are trained to analyze words, but students are visual animals. Well-conceived visualizations are not frills, they are foci for modes of thought. The course is driven by well-chosen, re-executable, interactive computer graphics and student-produced graphics inviting the students to experiment, to construct for themselves, to describe, and to explain what's happening in their own words.

Through interactive visualizations, Calculus&Mathematica tries to stick the basic calculus ideas into the students' unconscious minds before it transfers the ideas into English. For instance, students experiment with simultaneous plots of $f[x]$ and $f'[x]$ to acquire an understanding of the meaning of the derivative. Students experiment with plots of the exponential function and are imprinted with its awesome growth. Students who have never heard of convergence experiment with plots of functions and their Taylor series expansions, soon discovering that the convergence is what advanced mathematicians call "uniform on certain compact intervals."' And they invent the word "cohabitation" to describe what they see. Students experiment with running trajectories through vector fields and become comfortable with vector fields. They know that gradient fields drain at relative maximums. As a result of their experience, most of them can tell you why a solution of Laplace's equation cannot have an interior maximum. Reason: The gradient field of a solution of Laplace's equation has no sinks or sources.

Always give the students the opportunity for a creative response; give the students an active role in their own learning. Don't try to think for the students. Calculus&Mathematica students take an active role in their own learning by selecting material from the electronically alive Basics and Tutorials to learn (and possibly rework) as needed, and at their own pace. If a point doesn't get through, then they are free to modify and rerun as they see fit. At all times, they have the opportunity to pursue their learning actively and creatively. This lone aspect of C&M puts C&M at a great distance from lecture-based calculus courses and the new passive point-and-click multimedia courses coming onto the market. In the final analysis, this aspect of C&M is totally natural because this is the way research scientists do their work.

Approach mathematics as a science, not as a language or as a liturgy. Often mathematics is taught as a ritual or a liturgy in which the professor functions as curator of the dogma and arbiter of truth. Sometimes mathematics is taught as a language, a language which, as Blaise Pascal pointed out, "must be fixed in [the student's] memory because it means nothing to [the student's] intelligence." All too rarely is mathematics taught as the science that it is. The Calculus&Mathematica course attempts to teach mathematics as a science in which the student is the active investigator. With wise use of the computer to help introduce the ideas through the eyes, Calculus&Mathematica replaces the usual sequence,

lecture – memorization – tests,

with this variant of Mac Lane's sequence:

visualization – trial – error – speculation – explanation.

In this format, calculus becomes the same as the mathematical activity in which active mathematicians engage.

Ask students for explanations, not proofs. The words "prove" and "show" are the most terrifying words inexperienced math students ever encounter. The word "explain" is not so terrifying because explanations are usually assumed to be not so formal as a proof. On the other hand, a good explanation usually contains the main ideas of a formal proof; so that concentrating on explanations instead of formal proofs does not degrade mathematical understanding. In fact, rigor and understanding are often separate: Rigor is in one part of the brain, but understanding permeates the brain, the heart and the soul.

Rigor without understanding and understanding without rigor are both possible. In any case, the ability to recite a memorized proof of a theorem is not the same as understanding the theorem. The real goal is to understand. And that's what Phillip Davis is talking about when he says, "Classical proof must move over and share the educational stage and time with other means of arriving at mathematical evidence and knowledge."

Use a computer-based, genuinely interactive text. Conventional printed texts have a paralyzing effect on learning because they force the student into a passive, subservient role. Thomas S. Kuhn explains it best [3]: "Science students accept theories on the authority of teacher and text, not because of evidence. What alternatives have they, or what competence?"

The Calculus&Mathematica electronically alive interactive text, in which every example is as many examples as the student needs, is an environment in which the student can accumulate as much evidence as the student requires. The result: The student actively learns, in part, on the basis of the student's own authority and not just on the authority of the teacher or of the text.

Eliminate introductory lectures. Thurston [7] wrote: "Mathematicians have developed habits of communication that are often dysfunctional... most of the audience at an average colloquium talk gets little of value from it." Just as mathematics colloquium talks are usually failures, so introductory lectures in mathematics classes are usually failures. Reasons:

- Introductory lectures are full of answers to questions that have not been asked.
- By necessity, introductory lectures are full of precise terms not yet understood by the students.
- Introductory lectures provide the strong temptation for the teacher to try to do the thinking for the students.
- Introductory lectures tend to center the course on the lecturer instead of the students.

To paraphrase Schopenhauer: Attending introductory lectures is equivalent to thinking with someone else's head instead of with one's own. Instead of introductory lectures in Calculus&Mathematica, regular discussions are held, but not until the visual ideas have congealed in the students' minds as a result of their lab experience. These discussions emphasize answers to questions the students ask.

Motivate students to want to learn by serving up problems whose importance is recognized by the students. "What's this stuff good for?" is a question often heard from students in ordinary calculus courses but seldom heard from Calculus&Mathematica students. The reason is that the mix of student problems in C&M puts students in a position to try calculus out to see what calculus can do for them in terms of their own lives and in terms of their own planned professional futures in measurement, calculation and science. Students think carefully about how to apportion their efforts as part of their planned futures. Possessing an uncanny ability to recognize frivolous or artificial classroom problems, students usually tune out of ordinary calculus courses, but they rarely lose interest in Calculus&Mathematica.

Keep the language in the vernacular. Students fail in writing about mathematics because their textbooks are written in language that they cannot understand. As a result, they resort to rote memorization because much of what they read and hear means little to their intellects. Paul Halmos even went so far as to say that the job of the mathematics teacher is to translate the textbook into the vernacular. It does not have to be this way. Calculus&Mathematica is written in the vernacular in words, phrases and sentences that the students can understand and adapt in their own writing.

Give the students a chance to organize their thoughts by explaining themselves in writing. Calculus&Mathematica students visually absorb ideas uncorrupted by strange words, and they address the problem of communicating what they have learned only after they have a visual understanding of the idea under discussion. The first step is to visually determine what the truth is; the second step is to explain it. Students in ordinary calculus courses are deprived of the excitement of discovery and explanation. C&M students write a lot of mathematics and they are unexpectedly good at it. There are two reasons for this talent:

- The Mathematica Notebook front end gives the students a unified environment for graphics, calculations and write-ups.
- The language used in Calculus&Mathematica is informal enough for the students to adapt it to their own writing.

Give the students the opportunity to learn the mathematics and the programming in context. Ordinary attempts to bring applications into calculus tend to separate the mathematics from the applications. Similarly, ordinary attempts to bring technology into calculus tend to separate the mathematics from the technology. Calculus&Mathematica always puts the mathematics in the context of measurement and puts the programming in the context of mathematics. Most importantly, C&M exploits the technology in an effort to introduce new ideas.

As a result, the applications, the programming, and the mathematics all feed off each other. A C&M student put it best: "I have started to notice aspects of one class carrying over to another. Similarities in fields I thought unrelated before. An interconnection between math and language and programming and everything just kind of fits together a little better now."

Give the students professional tools. Students preparing for careers in a calculational science see computers or workstations running Mathematica as professional tools. Believing that the ability to use professional tools is part of their overall education, C&M students typically throw themselves into using Mathematica-equipped computers. They understand, perhaps better than their teachers, what vistas these professional tools open up.

Does it Work?

The study by Kyunmee Park and Kenneth Travers [6], which compares standard calculus and Calculus&Mathematica, states: "Generally the findings from an achievement test, concept maps, and interviews were all favorable to C&M students. The C&M group obtained a higher level of conceptual understanding than did the standard group without much loss of [hand] computational proficiency... [Some believe] that a laboratory course in calculus is very time consuming, and that students can become overly dependent on Mathematica. But this research found that the C&M course allowed the students to spend less time on computations and better direct themselves to conceptual understanding. Accordingly there was an increase in the students' conceptual achievement without a serious decrease in computational achievement... Furthermore, the C&M group's disposition toward mathematics and the computer was far more positive than that of the standard group... Generally, the C&M group seemed to more clearly understand the nature of the derivative and the integral than did the standard group... A positive side effect of the [computer] lab was the rapport that was established among the students. When students gathered around the computer, worked together, and shared and developed ideas, a great deal of mathematics was learned. ... [Computer] capabilities helped students discover and test mathematical results in much the same way that a physics or chemistry student uses the laboratory to discover and test scientific laws. Those capabilities provided the opportunities for the students to consider more open-ended questions and to encounter more realistic problems than often found in traditional calculus texts."

Students who enter calculus with high expectations and motivations resulting from their own professional plans in a calculational science are likely to blossom in C&M. This includes high percentages of engineering students and math students. It also includes motivated rural high school students in the C&M Distance Education Program at Illinois. Life science students at Illinois have done so well in C&M sections designed for life science students that the School of Life Sciences at Illinois has financed C&M labs for all of their freshman students.

We have been personally overwhelmed by the way students have thrown themselves into Calculus&Mathematica. We hope that Calculus&Mathematica and better courses to follow will help to pave the way to a time at which mathematics becomes just as alive for its students as it is for its practitioners.

The authors thank Paul Weichsel of the University of Illinois and John Ziebarth of the National Center for Supercomputer Applications for helpful comments.

References

[1] Bill Davis, Horacio Porta, and Jerry Uhl, *Calculus&Mathematica*, Addison-Wesley, 1994.

[2] Phillip J. Davis, "The rise, fall, and possible transfiguration of triangle geometry: a mini-history", *Amer. Math. Monthly* **102**:3 (March 1995), 204–214.

[3] Thomas S. Kuhn, *The structure of scientific revolutions*, University of Chicago Press, 1970.

[4] Saunders Mac Lane, "Responses to theoretical mathematics . . . ", *Bull. Amer. Math. Soc.* (*N.S.*) **30**:2 (April 1994), 190–193.

[5] *Moving beyond myths*, National Research Council Report, National Academy Press, 1991.

[6] Kyungmee Park and Kenneth J. Travers, "A comparative study of a standard and a computer-based college freshman calculus course", preprint.

[7] William P. Thurston, "On proof and progress in mathematics", *Bull. Amer. Math. Soc.* (*N.S.*) **30**:2 (April 1994), 161–177.

JERRY UHL
UNIVERSITY OF ILLINOIS
DEPARTMENT OF MATHEMATICS
1409 W GREEN ST.
URBANA, IL 61801-2917
UNITED STATES
   juhl@ncsa.uiuc.edu

WILLIAM DAVIS
OHIO STATE UNIVERSITY
DEPARTMENT OF MATHEMATICS
COLUMBUS, OH 43210-1328
UNITED STATES
   davis@math.ohio-state.edu

Contemporary Issues in Mathematics Education
MSRI Publications
Volume **36**, 1999

Japan: A Different Model of Mathematics Education

THOMAS W. JUDSON

Undergraduate and K–12 mathematics education in the United States have seen many reforms during the past decade. With the initial results of the Third International Mathematics and Science Study (TIMSS) appearing in October, 1996, there is also considerable interest in mathematics education elsewhere, especially Japan. A major revision in the K–12 curriculum in Japan has recently occurred, and the Mathematical Society of Japan has formed the Working Group for Undergraduate Mathematics, a committee of educators and mathematicians, to examine undergraduate mathematics education in Japan.

The success of the Japanese educational system in producing students who excel in mathematics is well-known and is pointed out in the results of TIMSS. This study was sponsored by the International Association for the Evaluation of Educational Achievement and involved approximately fifty nations world-wide. TIMSS focused on grades four, eight, and twelve, with Germany, Japan, and the United States receiving special attention [16; 17]. Unlike the National Council of Teachers of Mathematics *Standards*, which merely makes recommendations for K–12, Japan has a nationally set curriculum [5; 6; 12]. Since a fifth grade mathematics class in Tokyo will be covering roughly the same material as a class in Nagasaki or Sapporo during any given time of the academic year, Japanese educators have an opportunity to collaborate and polish lessons on a nationwide scale. This is not the case in U.S. schools, where the curriculum is locally controlled. In fact, the "TIMSS study of curricula found that current U.S. standards are unfocused and aimed at the lowest common denominator. In other words, they are a mile wide and an inch deep" [17]. In Japan, on the other hand, the achievements of students reflect the benefits of coherent goals and focused teaching practices. However, there is some question on how well the *Standards* have been implemented on a wide scale in the United States [17].

The purpose of this essay is to describe the Japanese educational system (with special emphasis on the mathematics component), and to compare and contrast it with that in the U.S.

K–12 Mathematics Education

In Japan, Munbusho, the Japanese Ministry of Education, sets the number of class periods for the year, the length of the class periods, the subjects that must be taught, and the content of each subject for every grade in K–12. For this reason, changes in the Japanese educational system are usually introduced more cautiously than in the United States, and possible curriculum revisions are evaluated more carefully before being put into effect. Technology-based courses of the type that one often sees in U.S. classrooms are not as popular in Japan, and Japanese educators generally seem to prefer a more traditional, theoretical, and problem-solving based course. Even though the current curriculum standards encourage the use of calculators beyond the fifth grade, calculators are still not allowed in many Japanese classrooms, since university entrance exams do not permit their use. Computers seem to be more prevalent in the Japanese classroom than hand-held technology [5; 6; 12; 13].

The elementary school curriculum is specified in Japan for grades 1–6. The goals and objectives of mathematics education at the elementary school level are to develop in children fundamental knowledge and skills with numbers and calculations, quantities and measurements, and basic geometric figures. In grades 1–3, children learn about the concept of numbers and how to represent them, the basic concepts of measurement, how to observe shapes of concrete objects and how to construct them, and how to arrange data and use mathematical expressions and graphs to express the sizes of quantities and investigate their mathematical relationships. They acquire an understanding of addition, subtraction, and multiplication, learn how to do basic calculations up to the multiplication and division of whole numbers, and learn how to apply these calculations. Children also become acquainted with decimal and common fractions during this time. The *soroban* or abacus is introduced in grade 3. Children learn basic concepts of measurement such as reading a clock, comparing quantities of length, area, and volume, and comparing sizes in terms of numbers. They are also taught the concepts of weight and time and shown how to measure fundamental quantities such as length [6].

By the end of grade 4, children are expected to have mastered the four basic operations with whole numbers and how to effectively apply them. They also should be able to do addition and subtraction of decimals and common fractions. In grades 5 and 6, children learn how to multiply and divide decimals and fractions. They are taught to understand the concept of area and how to measure the area of simple geometric figures and the size of an angle, as well as to understand plane and solid geometric figures, symmetry, congruence, and how to measure volumes. Children learn about the metric system during this time. Teachers show how to arrange data and use mathematical expressions and graphs to help children to become able to express the sizes of quantities. Letters such as x and a are introduced. Children also begin to learn about statistical data by using percentages and circle graphs (pie charts). It is recommended that calculators be introduced into the classroom in grade 5 to ease the computational burden [6].

Lower secondary school in Japan consists of grades 7–9. Preparation to get into the best high schools and universities begins at this time. There is tremendous pressure on students to perform well. Students are asked to learn a tremendous amount of material in grades 7–12, which is perhaps one of the major reasons why university and secondary school classrooms are often subdued. In contrast, elementary classrooms tend to be lively, with a great deal of interaction between students and teachers. In either case, classrooms are teacher-directed. The student-directed group learning that is found in some U.S. classrooms is virtually nonexistent in Japan.

In grade 7, students learn about positive and negative numbers, the meaning of equations, letters as symbols, and algebraic expressions. By the end of grade 8, they are able to compute and transform algebraic expressions using letter symbols and to solve linear equalities and simultaneous equations; they have also been introduced to linear functions, simple polynomials, linear inequalities, plane geometry, and scientific notation. In grade 9, students learn how to solve quadratic equations (those with real solutions) and are taught the properties of right triangles and circles, functions, and probability. In grade 7 and beyond it is recommended that calculators and computers should be efficiently used as the occasion demands [6].

In high school (grades 10–12), six mathematics courses are offered: Mathematics I, II, III and Mathematics A, B, and C. Although only Mathematics I is required of all students, those students intending to enter a university will usually take all six courses. In fact, Japanese high school students who take all of the courses offered will know more mathematics than many U.S. students do when they graduate from college. In Mathematics I, students are taught quadratic functions, trigonometric ratios, sequences, permutations and combinations, and probability. Mathematics II covers exponential functions, trigonometric functions, analytic geometry (equations of lines and circles), as well as the ideas of limits, derivatives, and the definite integral. Calculus is taught in Mathematics III, including functions and limits, sequences and geometric series, differential and integral calculus. More advanced topics such as Taylor series are usually not taught in Mathematics III. Mathematics A deals with numbers and algebraic expressions, equalities and inequalities, plane geometry, sequences, mathematical induction, and the binomial theorem. Computation and how to use the computer are also taught in this course. In Mathematics B, students learn about vectors in the plane and 3-space, complex numbers and the complex number plane, probability distributions, and algorithms. Mathematics C consists of a variety of topics, including matrix arithmetic (up to 3×3 matrices), systems of linear equations and their representation and solution using matrices, conic sections, parametric representation and polar coordinates, numerical computation including the approximate solution of equations and numerical integration, and some calculus-based statistics [6; 20].

The University Entrance Exams

The importance of university entrance exams in Japan cannot be overstated, since admission to the "right" university may dictate one's future career and social status. From the time that a student enters lower secondary school, much of the Japanese educational system is dedicated to preparing students to pass the university entrance exams. Throughout high school (and before) students often attend *juku* and *yobiko* after regular school hours or during holidays. These are special cram schools that prepare students for the university entrance exams.

The University Entrance Center Examination (UECE) is similar to the SAT exam given in the United States. The entrance exam to public universities consists of two parts. The standardized primary exam, the first part, is offered once per year in mid-January. This exam is administered by Munbusho. The secondary exam is offered by each university at a later date. Usually the exams are weighted about 50–50; however, at some prestigious universities the secondary exam is given more weight. The University of Tokyo counts the secondary exam as 80%. Private universities will either use the UECE or use their own exam. A small percentage of students, usually at private universities, do not have to take the examinations and are admitted by recommendation [19].

University Mathematics Education

In Japan, the school year begins in April and ends the following March. Most universities are on the semester system; however, a few institutions use the quarter system. The first semester runs from the beginning of April until mid or late July. The second semester begins in mid-September and ends in early February. Undergraduate classes in Japan tend to meet less often than those in the United States. While students in a freshman calculus class in the U.S. meet three to five times a week, it is not uncommon for the same class in Japan to meet only once a week for a 90 minute period and cover the same amount of material.

We often hear that students in Japan work very hard in high school and that their time at the university is a four-year vacation. This may be true at many universities and in certain disciplines; however, there are other universities that demand that their students work hard. Those who choose to major in disciplines such as engineering and medicine are very serious students.

Although the undergraduate mathematics major in Japan is perhaps more theoretically oriented, the course of study is similar to what we see in the United States. The actual curriculum varies with each university, but the first three semesters typically consist of a repeat of single variable calculus followed by multi-variable calculus. The undergraduate calculus course is much more rigorous than the high school course. At the upper level, students are expected (but not necessarily required) to take the usual courses in linear algebra, real and complex analysis, geometry and general topology, and algebra. Often a thesis or a senior seminar is required. Technology in the classroom seems to be more prevalent at the university undergraduate level than in K–12; however, graphing calculators and computer algebra systems are nowhere near as popular as they are in the United States [1; 3; 13; 20].

Like many U.S. professors, university professors in Japan are not satisfied with their incoming students. In a recent Mathematical Society of Japan survey, 80% of the university professors surveyed felt that the mathematical ability of incoming students has declined. Between November 1995 and January 1996, approximately 150 teachers were surveyed at public and private universities. Of the 84 responses received, 78% felt that the mathematical abilities of university students had declined, and no one indicated that the situation had improved. The majority of the respondents felt that abstract, logical, and mathematical thinking ability had declined. The responses also indicated that foundational skills, the ability to read and understand Japanese, and the ability to apply mathematics had deteriorated. Some respondents also felt that the ability to calculate had declined and that there was greater student apathy. Responses to the survey indicated that the problem was K–12 education and the university entrance exam system. They felt that there was a decrease in the amount of time spent in the classroom on mathematics and that the memorization model of study for the university entrance exams was a detriment to learning abstract thinking and problem-solving skills [18].

Considerable collaboration between Japan and the United States in K–12 mathematics education has occurred; but similar cooperation is only beginning to materialize at the calculus level and beyond [2; 3; 4; 13; 15; 20]. There is also a growing interest in using technology to teach mathematics in Japan and other Asian countries. The First Asian Technology Conference in Mathematics, held in Singapore in December, 1995, was well-attended by mathematicians and educators from Japan, other Asian-Pacific countries, Europe, and the United States [1]. The next two ATCM conferences are scheduled for June, 1997 in Penang, Malaysia and August, and 1998 in Tsukuba, Japan. In 2000, the Ninth International Congress of Mathematics Education will be held in Chiba, Japan.

References

[1] Association of Mathematics Educators. *Proceedings of the First Asian Technology Conference in Mathematics.* Singapore: Association of Mathematics Educators, 1995.

[2] Jerry P. Becker, editor. *Report of U.S.-Japan Cross National Research on Students Problem Solving Behavior.* Carbondale: Department of Curriculum, Instruction and Media, Southern Illinois University, 1992.

[3] Jerry P. Becker and Tatsuro Miwa, editors. *Proceedings of the United States-Japan Seminar on Computer Use in School Mathematics.* Carbondale: Department of Curriculum, Instruction and Media, Southern Illinois University, 1992.

[4] Illinois Council of Teachers of Mathematics. *Mathematics teaching in japanese elementary and secondary schools: a report of the ICTM Japan Mathematics Delegation (1988).* Carbondale: Southern Illinois University, 1989.

[5] Ishizaka, Kazuo. *School education in Japan.* Tokyo: International Society for Educational Information, Inc., 1994.

[6] Japan Society of Mathematical Education. *Mathematics program in Japan (kindergarten to upper secondary school).* Tokyo: Japan Society of Mathematical Education, January, 1990.

[7] Kodaira, Kunihiko, editor. *Algebra and geometry: Japanese grade 11.* Providence: American Mathematical Society, 1996.

[8] Kodaira, Kunihiko, editor. *Basic analysis: Japanese grade 10.* Providence: American Mathematical Society, 1996.

[9] Kodaira, Kunihiko, editor. *Mathematics 1: Japanese grade 10.* Providence: American Mathematical Society, 1996.

[10] Kodaira, Kunihiko, editor. *Mathematics 2: Japanese grade 11.* Providence: American Mathematical Society, 1996.

[11] Curtis C. McKnight et al. *The underachieving curriculum: assessing U.S. school mathematics from an international perspective.* Champaign, IL: Stipes Publishing Co., 1987.

[12] National Council of Teachers of Mathematics. *Curriculum and evaluation standards for school mathematics.* Reston, VA: National Council of Teachers of Mathematics, 1989.

[13] Robert E. Reys, and Nobuhiko Nohda, eds. *Computational alternatives for the twenty-first century: cross-cultural perspectives for Japan and the United States.* Reston, VA: National Council of Teachers of Mathematics, 1994.

[14] Sato, Teiichi. *Higher education in Japan.* Tokyo: Ministry of Education, Science, and Culture, 1991.

[15] James W. Stigler and Harold W. Stevenson. "How Asian teachers polish each lesson to perfection," *American Educator: The Professional Journal of the American Federation of Teachers* 15.1 (1991): 12+.

[16] *Third International Mathematics and Science Study* (6 Dec. 1996): Internet. January 2, 1997. See http://www.ed.gov/NCES/timss/brochure.html.

[17] *US TIMSS* (6 Dec. 1996): Internet. January 2, 1997. See http://ustimss.msu.edu/.

[18]   *Working Group for Undergraduate Mathematics, Mathematical Society of Japan* (in Japanese). (17 Dec. 1996): Internet. January 2, 1997. See http:// skk.math.hc.keio.ac.jp/mathsoc/wg-homepage.html.

[19]   Ling-Erl Eileen T. Wu. *Japanese university entrance examination problems in mathematics.* Washington: Mathematical Association of America, 1993.

[20]   Zhang, Dian-zhou, Toshio Sawada, and Jerry P. Becker, editors. *Proceedings of China-Japan-U.S. Seminar on Mathematics Education.* Carbondale: Department of Curriculum, Instruction and Media, Southern Illinois University, 1996.

THOMAS W. JUDSON
DEPARTMENT OF MATHEMATICS AND COMPUTER SCIENCE
UNIVERSITY OF PORTLAND
5000 N WILLAMETTE BOULEVARD
PORTLAND, OR 97203-5798
UNITED STATES
   judson@uofport.edu

The Debate over School Mathematics Education

Contemporary Issues in Mathematics Education
MSRI Publications
Volume **36**, 1999

Reflections on Teacher Education

ANNELI LAX

Prologue

Our panel's topic is one of the hotly discussed national issues: K–12 mathematics instruction. Mathematics is an important thinking tool, being

- a way of looking at and making sense of the world;
- a way of, for example, checking out the validity of claims made by advertisers or politicians;
- a way of reasoning.

Certain mental habits, observable in young children, should be cultivated in school rather than inhibited or allowed to atrophy.

Mathematics is particularly well suited for exercising mental muscles. Even those who do not achieve virtuosity can find that the psychological rewards of feeling mentally fit and gaining control of one's reasoning powers are just as great as the rewards of physical fitness and control of one's body.

Classroom teachers, always considered essential in the education of our children, are now expected to play even more crucial roles. Reformers eager to save public education see teachers as "agents of change". Advocates of children who see so many deprived of functioning families want teachers to act in *loco parentis*. Future employers have discovered that they no longer need human robots but flexible thinkers who can work in teams to solve problems with tools appropriate to the task. They are joining educators (who want to build communities of future adults able to cope with the demands of the next century) in viewing teachers as coaches in the acquisition of skills, and as guides in discussions of conjectures and verifications of results. Our demands on teachers are overwhelming indeed.

I want to talk about the role mathematicians might play vis-à-vis schools and teachers. If time permits, I should like to make this discussion less theoretical by telling anecdotes from my personal experience in some inner city schools of New York City, observations in an "alternative school", regular visits to planning sessions of mathematics and science teachers in a recently created "small" public

school, and monthly "math suppers" with about a dozen dedicated teachers
who work in various schools: private, suburban, alternative, a newly created
secondary school modeled after Central Park East, "Coalition" schools, and a
community college.[1]

Many teachers are confronted by the conflict of wanting to teach a curriculum
that they consider to be right for their students, and on the other hand feeling
that they must help students pass tests that they consider irrelevant, but that
afford entry to certain careers and/or future studies. Mathematicians may help
resolve this conflict.

How Can Mathematicians Help
Mathematics Instruction in K–12?

University and industry research mathematicians can make a difference; look
at Leon Henkin, Paul Sally, Arnold Ross, Henry Pollak, or, among the next
generation, Phil Wagreich, Herb Clemens, Judy Roitman and others. Here are
two ways they can help:

- By working, as almost silent partners, with teachers on planning classes and
  seeing how the plan works, discussing it afterwards with the teachers (perhaps
  also the students), and by improving the strategies for the next round of
  teaching that topic.

For most mathematicians, getting into contact with classroom teachers in their
schools takes time, patience, and luck. We have to take down barriers. Educators say "What do these ivory tower mathematicians know about adolescents
and about pressures from districts, parents, administrators?"; and university
mathematicians say: "Look what these people are doing to our beautiful subject! And to some of our brightest students!" Teachers are often overburdened
not only with too many classes, too many students per class, but also with non-
instructional chores and bureaucratic paperwork. They cannot find the time to
get together with mathematicians, and often are afraid of being found wanting
mathematically. Mathematicians feel they are not trained to teach children and
should do that which they are good at and which is rewarded: research. It took
me more than a decade to be welcomed in schools.

Seeing teachers in action made me admire them, especially for those talents
they have and I lack, such as engaging groups of kids in tasks they perform
collaboratively. But I have found most teachers weak in mathematics. Many
enjoy what I can offer, in planning sessions outside their classrooms, particularly

---

[1]Time did not permit. But I was inspired by Dick Askey's careful comments in his article
(published in this volume) to use the same mathematical example to fashion a strategy richer
in mathematical and pedagogical opportunities than a mere enumeration of weaknesses in the
text. I called the paper *When you find a lemon, make lemonade!* It is included as an appendix
to this paper; see page 90.

my connecting a student question or error to some part of mathematics or to an application or generalization or to a test item. One can go quite deep (but in small doses), and one must follow through. A great part of learning involves connecting a new piece of knowledge to something you already know or at least vaguely remember. But you need to have something in your experience to which you can connect the new event. Mathematicians have lots of such things in their heads, teachers have some, and students have only a few and need to build more. Nobody can say "this reminds me of a joke" if he has never heard that joke. Learning to make connections is one of the most important skills to emphasize in school.

- By looking critically, with teachers, at materials (textbooks, manipulatives, videotapes) adopted or recommended by their schools or by regional or national committees; by getting teachers to suggest what to emphasize, what to omit, and why, and how to correct or supplement weak sections of the text; by visiting a group of teachers who are creating a new curriculum unit. There are now some good texts that are making a difference.

Good teachers, at any level, rarely follow a textbook faithfully, even if they have authored it. In selecting a text, then modifying it, we use our mathematical knowledge as well as our classroom experience. By listening to students and reading their assignments, we become aware of what students find difficult and why; we note where misconceptions and anxieties are formed; our mathematical knowledge together with our curriculum agenda combine to lead students into meaningful mathematics.

I cannot think of better "staff development" than critiquing, correcting, supplementing and trying out a curricular unit. This should become a regular part of teacher education in collaborative set-ups where at least one non-obtrusive mathematician takes part. The success of such efforts depends on the personalities and interests of the participants. The arts of connection-making and of salvaging the useful parts of errors help, and the fear of exposing gaps in knowledge can get in the way.

Interested mathematicians are likely to find some obstacles, such as the barriers described above, and not being invited to give feedback on preliminary versions of a new curriculum, either because a publisher co-sponsors it and has a copyright, or because the creators fear their material may be misused by untrained teachers and cause more harm than good. It may even happen that research mathematicians are asked to serve on advisory committees to curriculum projects and find themselves "token mathematicians", neither consulted nor heeded. After they resign from the advisory committee, their names keep appearing on the list of members.

New and Old Styles of Teaching

Some years ago I saw a television program that featured two teachers: The first was shown lecturing to a class in a most engaging and stimulating way; the second was shown asking some challenging questions, getting groups of 4 or 5 to work together, walking around the class room stopping occasionally to clarify or ask a student to clarify a point that arose, and demanding a verbal description from a member of each group of what results or partial results had been arrived at. I was delighted by the performance of each and impressed by their knowledge of the topics studied in each class. It would never occur to me to persuade either one of these teachers to adopt the methods of the other, although I believe more young students benefit from the latter model because of their active engagement and intellectual struggles and the necessity of talking and listening to other group members who may bring a different perspective to the problem under consideration.

Let us encourage diversity in teaching styles even as we take into account diversity in learning styles. Different styles will reach different children; let all experience learning in more than one way. Let us not cramp a teacher's style! If the teacher is coerced into a mode felt to be unnatural, he or she will become less effective.

Reformers as well as traditionalists want students to reason mathematically, to express their thoughts and to examine one another's methods of attacking and solving a problem. Lecturers can reveal their own more informed thoughts in presenting a solution, or they can show students the stunningly clever thoughts of some great mathematicians by lecturing about their ideas. The pitfall to avoid is to have the class say "I could never have thought of any of these things; I might as well give up." It is the second teacher who can better demonstrate to students that, yes, they can indeed think of some pretty clever things. University mathematicians do just this with their doctoral candidates and should keep it in mind when they think of a young novice. The main difference is that the doctoral student must think of something completely new, while young students are discovering something well-known, yet new to them. Both teachers and students work and learn best when they feel that they are doing their own work and not just following somebody else's unexplained prescriptions.

The "basic skills" people should stop vilifying the "concepts" people by characterizing them as sentimental "hand holders" who lack appreciation of mathematical rigor; and the "concepts" people should stop thinking of the "skills" people as mindless drill masters without either aesthetic or deductive tendencies. Clearly, mathematical activities require both: understanding of concepts and facility with technique. But we should realize that some people first learn rules and algorithms and then become interested in why these work so well, while others are unable to accept recipes unless their origins and logic are explained. Most of us combine the two modes.

What Does all This Have to do with Teacher Education?

Both teachers shown in the TV show mentioned at the beginning of the previous section were well grounded in mathematics. I don't know how many courses or credits each had, but they knew that mathematics is not just a collection of "facts", and they recognized that learning mathematics involves doing mathematics. In contrast, the well worn caricatures of the traditional drill master (who concentrates on endless basics) and of the overzealous "facilitator" (who neglects proofs and questions of structure) would, with better mathematical background, become more like the models I saw on TV.

I want to cite the work of two people: Deborah Ball [Ba] asks "Not How Much, but What Kind?" of mathematics should teachers know? [Ba] And Eileen Fernandez, in a work in progress [Fe], asks "How does a teacher's knowledge affect her response to unanticipated student questions?"

The first of these focuses on the irrelevance of most quantitative certification requirements, e.g. 12 credits of mathematics courses above College Algebra; instead she examines what teachers make of a taped student discussion, how they interpret what is going on in students' heads as they argue about whether or not 6 is an even number, and what mathematical knowledge would help them guide the students in fruitful directions.

The second of these, also concerned with what students are thinking (when they are not thinking what we think they ought to be thinking) examines teacher responses to what students are saying, what errors the students are making, where they get stuck. She will probably conclude that the teachers' responses depend very much on the connections these teachers make to things they know, to mathematics they have internalized. Both researchers evidently consider teachers' mathematical strengths crucial because of their role in teacher-student interactions and in finding fruitful paths out of student errors and into important domains not necessarily next in the syllabus.

As college instructors of prospective teachers, we should give students an opportunity to become engaged in the kind of explorations and collaborations that many educational reformers advocate, yet do not provide for the practitioners. We, as well as the teachers we educate, should stop making assumptions about who learns in what ways. Some claim children are incapable of abstraction or generalization—they are wrong. Some look with disdain on the utilitarian "real world" views of mathematics. They too are wrong. What is needed is a balance, and if some of us have biases in one direction, let these be counterbalanced by colleagues of different opinions; but let us all try to take our students' talents and interests into account as we develop their mathematical teaching crafts.

In conclusion, I want to thank my panel colleagues for providing balance to the raging controversies on reform. I was cheered by George Andrews's article [An], and by David Mathews's response and Andrews's rejoinder in the same volume. There have been many thoughtful articles and reports on significant educational insights. Many of these preceded the reports we hear and read about. Yet, when you visit most class rooms, you see no change (except perhaps more anxiety on the part of teachers and administrators about meeting new standards or preparing students for changes in tests). Similarly, we see little change (with some notable exceptions) in the practices of most teacher education departments, and we see almost no change in the mathematics courses or seminars that Liberal Arts colleges recommend to prospective teachers. We need strong consortia of collaboratives to give prospective teachers the pedagogical and mathematical experiences that would prepare them best for what they would like and what we expect them to do. It is up to mathematicians to take such initiatives.

Appendix: When you Find a Lemon, Make Lemonade!

A recent text contains the problem:

Explain why $\sqrt{4}$ is rational while $\sqrt{5}$ is irrational.

The teachers' edition of this text suggests the answer:

"$\sqrt{4} = 2$ which is rational; $\sqrt{5}$, in its decimal form, does not terminate or repeat and therefore cannot be written as an integer over an integer."

Now we don't know what the class has studied prior to being given this problem; presumably they know the definition of "rational" and, my guess is that they have seen a proof of the irrationality of $\sqrt{2}$, in which case they might try to cook up a similar proof for $\sqrt{5}$. But now suppose the teacher tells his class what the teachers' edition says. Two questions would immediately arise:

1. How do you know that the decimal form of $\sqrt{5}$ does not repeat or terminate?

2. Even if you knew that, how would the irrationality of $\sqrt{5}$ follow from the fact that its decimal representation is infinite and nonrepeating?

This would lead to two discussions which may be carried on with the entire class; or two groups might be formed, each pondering one of these questions. About Question 1, the following conversation might take place:

STUDENT A: When I put 5 into my calculator and press the $\sqrt{\ }$ button, I get 2.2860679775, and this is only 10 places. This does not tell us if there will be a repeat, only that we don't see one so far.

STUDENT B: My sister is in college and can get something called "double digit accuracy" on her computer.

STUDENT C: So what! Suppose you get twenty or even 1024 places, which, they tell me, some computers can get, you still would not know if, may be after a million places, some string of digits starts repeating.

STUDENT A: I remember that my grandmother told me she learned a way of calculating square roots in school long before there were computers or calculators, and that this was even harder than long division; but you could get as many decimals as you want.

STUDENT C: And if you sat there all your life, you still wouldn't know if digits might repeat eventually.

At this point the class would conclude that the advice they got is not very helpful, because there is no way of verifying the claim that the decimal expansion neither repeats nor terminates. Here, we might hear a subdiscussion where A promises to ask his grandmother to show him the method she learned, and somebody else volunteering an uncle who knows how to extract the square root of a number N by first guessing a nearby integer, say x_0, and then refining the guess by averaging x_0 and N/x_0 and using this as the next approximation x_1, and then continuing this process, getting successive approximations $x_n = \frac{1}{2}(x_{n-1} + N/x_{n-1})$. At a future class meeting, the uncle's and grandmother's methods may even be compared for efficiency and accuracy; for example, the students can determine after how many iterations in the uncle's method they get the 10 digits the computer gave them, etc.

Now let us turn to Question 2. Suppose we knew by some feat of omniscience that $\sqrt{5}$ has an infinite, nonrepeating decimal representation. Then we may hear

STUDENT C: Well, if we knew that all rational numbers have terminating or repeating decimal expansions, then we could conclude that a number whose decimal expansion neither repeats nor terminates must be irrational simply because, if it were rational, this could not happen.

STUDENT D: Oh yeah! We had examples like this in the school I went to last year, in a unit on Logic. Some fancy name was used, like 'contrapositive' I think. Well, how would you prove that rational numbers have repeating or terminating decimal expansions?

STUDENT E: Well, how do you find the decimal representation of a number that can be written in the form $\frac{a}{b}$ with a and b whole numbers, $b \neq 0$? You would divide a by b.

And here comes a good reason for studying the division algorithm instead of using the calculator. In this discussion, students would probably take some examples, perhaps 2/3 or 1/7 or 3/5 or some other fraction with a small denominator and would be led to some notable observations. For example, if a fraction in lowest terms has a denominator with no prime factors except 2 and 5, then its decimal representation must terminate. This will reinforce the notion that a terminating decimal is merely a fraction whose denominator is a power of 10. In the case 3/5 above, we merely multiply top and bottom by 2 and get $3 \times 2/5 \times 2$ or $6/10 = 0.6$. In the other examples, they would see that in the long division, remainders keep popping up, and since the only nonzero ones are $1, 2, \ldots, b-1$ (where b is the denominator), we see by the pigeonhole principle that one of these remainders must, after at most $b-1$ steps, pop up again. And when this happens, the whole process repeats and we get a periodic decimal expansion.

The discussions of questions 1 and 2 need to be merged and the conclusions summarized. Yes, the decimal representation of rational numbers is periodic or terminating, so a number with infinite, nonperiodic decimal representation is indeed irrational; the trouble is we cannot tell that $\sqrt{5}$ is in this category, so we need to figure out another way of proving its irrationality.

We may now recall (or look for the first time at) the proof of the irrationality of $\sqrt{2}$ and try to construct its analogue for $\sqrt{5}$—and more generally for $\sqrt{M}$, where the integer M is not a perfect square—and probe a bit into elementary number theory. We may also recall that an argument for the existence of $\sqrt{2}$ on the number line was the fact that it has geometric representations, for example as the length of the diagonal of a unit square, and we come up with the geometric analogue: $\sqrt{5}$ is the length of the diagonal of a 1×2 rectangle. Both will reinforce the theorem of Pythagoras. Students who like geometry may find other ways of constructing $\sqrt{5}$ and other square roots.

Another outcome may be a more precise application of the very elementary and extremely useful pigeonhole principle. It shows, for example, that the period of the decimal expansion of a rational number in lowest terms cannot be longer than the size of its denominator. So if the computer output of 2^{10} digits shows no periodicity, we could conclude that, if $\sqrt{5}$ were rational, its denominator would have to be at least 1023. Though the pigeonhole principle is understandable to everybody, the difficulty in applying it is often due to the difficulty we have in deciding which are the objects to be stashed, and which are the pigeonholes that must hold these objects.

Computer buffs in the class may enjoy writing a program for the iteration scheme that the hypothetical uncle has contributed, see above.

I am not suggesting that all these connections will be made; but I do want to call attention to the benefits of analyzing a suggested method for solving a problem. We can learn a lot even if the method cannot be carried out in finite time. We may also gain a better appreciation for an easily applicable, efficient, and correct method for solving our problem.

References

[An] G. Andrews, "A case for balance", *The College Math. J.*, November 1996.

[Ba]   D. Ball, "Unlearning how to teach mathematics", in *For the Learning of Mathematics*, C. Gattegno. Reading, Eng., Educational Explorers Ltd., 1963.

[Fe] E. Fernandez, "The standards-like role of teachers' mathematical knowledge in responding to students' unanticipated perspectives", paper presented at the 1997 Annual AERA Conference, Chicago.

ANNELI LAX
COURANT INSTITUTE
NEW YORK UNIVERSITY
251 MERCER STREET
NEW YORK, NY 10012-1110
UNITED STATES
   al4@scires.nyu.edu

Contemporary Issues in Mathematics Education
MSRI Publications
Volume **36**, 1999

The Third Mathematics Education Revolution

RICHARD ASKEY

Introduction

The three mathematics education revolutions in my life were called the New
Math, the reaction to it called "Back to Basics", and the current one, which is
best illustrated by the NCTM Standards.

In the New Math period, many mathematicians were active in what would
now be called education reform. However, there were also some mathematicians
who expressed serious reservations about what was happening. Morris Kline
was the most vocal, but far from the only one. A group of 64 mathematicians
signed an article protesting the direction taken by the then current reforms. This
article, "On the mathematics curriculum of the high school", was published in
both the *American Mathematical Monthly* and in *The Mathematics Teacher* [1].
It was also reprinted in Kline's book *Why Johnny Can't Add* [16]. While the
problems associated with the New Math were different from those the current
reforms are causing, this article is still worth reading. Here is one paragraph
from it:

> 6. 'Traditional' mathematics. The teaching of mathematics in the ele-
> mentary and secondary schools lags far behind present day requirements
> and highly needs essential improvement: we emphatically subscribe to this
> almost universally accepted opinion. Yet the often heard assertion that
> the subject matter taught in the secondary schools is obsolete should be
> closely scrutinized and should not be taken simply at face value. Elemen-
> tary algebra, plane and solid geometry, trigonometry, analytic geometry
> and the calculus are still fundamental, as they were 50 or 100 years ago:
> future users of mathematics must learn all these subjects whether they are
> preparing to become mathematicians, physical scientists, social scientists
> or engineers, and all these subjects can offer cultural values to the general
> students. The traditional high school curriculum comprises all these sub-
> jects, except calculus, to some extent; to drop any one of them would be
> disastrous.

Contrast this with what is being written now:

> California has taken the lead in upgrading mathematics education for the
> 21st century. But while reforms are gradually taking hold, the majority
> of classrooms still rely on a traditional mathematics curriculum, that, as
> one cynical observer remarked, is largely composed of eight years of 15th
> century arithmetic, two years of 17th century algebra and one year of 3rd
> century B.C. geometry. [21]

It is not clear what Calvin Moore meant when he quoted a cynical observer,
but it should not mean that we should drop any of these topics. When the
arithmetic developed in India passed through the Arab Near East to Italy, the
result was the flowering of mathematics that not long afterwards led to the
development of algebraic tools by Viète and others in the 16th century and then
to the analytic geometry of Descartes and Fermat. This blending of algebra
and geometry enriched both subjects, and led to the development of calculus.
It is surprising that the NCTM Standards both pushes for the interconnection
between different parts of mathematics and yet calls for less emphasis on conic
sections [23, p. 127]. The algebraic treatment of conics is an ideal place to show
how algebra and geometry can be blended, with both enriched as a result. The
texts written before [23] was published in 1989 contained little about conics —
just their definitions via a focus or foci and distance, such as the sum of the
distance from the two foci being constant for an ellipse, the derivation of standard
equations for them from the definitions, and the reduction to this standard form
when the curve is translated. Translation of curves is now a major part of how
graphing calculators are used in algebra. This leaves nothing to decrease without
removing everything else. That is a pity, since what is dropped is a major part
of the essential connection between geometry and algebra.

Similarly, Greek geometry played a vital role in the development of mathematics. While mathematics of some sort seems to have been developed in every
culture, the idea that simpler parts can be used in a systematic way to develop
more complicated parts was not wide-spread. For example, Chinese mathematics had an almost exclusively practical focus. Greek geometry reached China
through the missionary Matteo Ricci in the first part of the 17th century. He
and Xu Guanggi translated the first 6 books of Euclid's elements around 1607.
Xu expressed great admiration for the ideas in this work. Here is a sample.

> [I]t looks complicated, in fact it is supremely simple, so we can use its
> simplicity to simplify the complexity of other thing; it looks difficult, in
> fact it is supremely easy, so we can use its ease to ease other difficulties.
> Ease arises from simplicity and simplicity from clarity; finally, its ingenuity
> lies in its clarity.

This is taken from [18]. For a view from India, consider the book review [24] of
[14].

I would like to end with a few comments on this thought-provoking and informative book. It is true that all the civilizations of the past have thought about questions in arithmetic. But it cannot be denied that modern mathematics, as it is understood today, does owe a great deal to the renaissance in Europe, which in turn was a miraculous revival of Greek thought.

These classical parts of mathematics are still of great use, and need to be taught and learned.

The "lead" in mathematics education which Moore wrote about California having taken comes from their 1992 Mathematics Framework [3]. In October, 1996, Wisconsin was in the process of trying to write State Standards. The Wisconsin Academy of Sciences, Arts and Letters appointed a committee to look at the Mathematics Standards and comment on them. At one meeting, I passed out a copy of the newly released California Mathematics Program Advisory [4] and said this seemed to be the first step in replacing this 1992 Framework. Tom Romberg, who chaired the committee which wrote the NCTM Curriculum and Evaluation Standards, replied: "About time". At the annual meeting of the National Council of Teachers of Mathematics in April, 1997, Romberg gave a talk in which he singled out these Frameworks as distorted, focusing almost exclusively on pedagogy instead of dealing with mathematical content. He said that he had written a warning letter to the people who were writing these Frameworks when it was in draft stage. This talk was based on a monograph Romberg is writing. The talk was titled "Sometimes it is frightening when people take you seriously: Reflections on the Standards".

Of the subjects listed in the paragraph quoted from [1], solid geometry has almost completely disappeared from the high school curriculum. Analytic geometry has disappeared with the exception of straight lines and circles. Due to the heavy reliance on graphing calculators, the only form of a straight line most students in calculus know from high school is $y = mx + b$. When asked to find the equation of a line through two points or the equation of a line with a given slope but a given point off the y-axis, most students have no idea what to do. The new standard text definition of a parabola is the equation

$$y = ax^2.$$

No geometry is mentioned.

If there were nothing more to the latest revolution in mathematics education than statements like the one made by Moore, we could probably relax and say this will pass. However, textbooks have been written to "conform" to the NCTM Standards and some of them are sorely lacking what is needed to improve mathematics education. There is a lot of rhetoric about teaching for understanding, yet here are some examples of what we get.

A Case Study

The Addison-Wesley book *Focus on Algebra* [5] was described but not named in an article which appeared in the Christian Science Monitor under the heading " 'Rain Forest' Algebra Course Teaches Everything but Algebra" [13]. The author of the article, Marianne Jennings, is a professor at Arizona State University and had a daughter in a class which used this book as a text. The daughter was getting an A in beginning algebra, but had no idea how to solve an equation. We can look at the book and see that equations are not introduced until page 165, and are first considered as functions. The first solution of a linear equation comes on page 218, and is by guessing and checking. Next, graphs are used to solve them. Then come algebra tiles. If you have never seen algebra tiles, let me strongly suggest you find a relatively new first year algebra book and read about them. They will be described a bit later, but you really have to see them in full color to appreciate what is happening in school algebra. Finally, on page 255, a method is introduced which has some "power", something which is written about regularly in reform circles but is seldom encountered.

Ms. Jennings was not convinced that any algebra was being done until her daughter could solve

$$3x = x + 4$$

by subtracting x from both sides and then dividing both sides by 2 to get $x = 2$. One of the many problems in the New Math was that parents did not understand what was being done or why. Ms. Jennings knew her daughter could not solve the equation above by what she thought was a simple method, so went to school to ask what was happening. The daughter's teacher said: "We don't plug and chug anymore. We're teaching them to think." The teacher assured her that in five years her daughter and the other kids would be great in math.

I lived through the New Math as a parent, and was immediately concerned by the response of the teacher. I had heard all of this before, so called Ms. Jennings to find out which book her daughter's class was using. After finding a copy in a local high school, I looked at it. Some of my concerns follow.

The book is over 800 pages long. In Howson's book [12] on textbooks written for the TIMSS study, he points out that US textbooks are much longer than those in the other countries whose books he looked at. If more significant material were covered than in other countries, or if the explanations were better, the length might not be a problem. Unfortunately, neither is the case.

Lack of Reasoning

To see some of the failings of *Focus on Algebra*, consider the comment made by the teacher, "we teach our students to think". In the review section of the last chapter, there are some questions dealing with irrational numbers. One is to explain why $\sqrt{4}$ is rational and why $\sqrt{5}$ is irrational. In the teacher's edition, the answers are $\sqrt{4} = 2$ which is rational and $\sqrt{5}$, in its decimal form, does not terminate or repeat and therefore cannot be written as an integer over an integer. The teacher who thought this book was teaching students to think might think that this is an example of thinking, but there is no one in the world who knows how to prove that the decimal expansion of $\sqrt{5}$ does not repeat without first showing that it is not rational.

The substitution of rote comments that are misleading for substance should be contrasted with the treatment of irrationality of square roots in both a Japanese text [17] and in the book written for the Gelfand Outreach Program [9]. Each of these contains a proof of the irrationality of the square root of 2. With this knowledge, it would be appropriate to ask the students to explain why square root of 5 is irrational.

The last chapter is titled "Functions and the Structure of Algebra". It starts with a brief comment about the golden rectangle and the golden ratio, but without mentioning similarity. Students are asked why the golden rectangle might be more attractive than other rectangles. A possible response which is suggested is that the shapes are not too long and not too square. At this point, the golden ratio is said to be a number approximately equal to 1.618. In the first section, it is said to be the number

$$g = (1 + \sqrt{5})/2$$

which is approximately 1.618034. Then the authors move immediately to the Fibonacci sequence. The word "sequence" does not appear in the index, nor are any other sequences mentioned there, so one wonders why these are not given the traditional name of Fibonacci numbers. The first seven Fibonacci numbers are stated and the rule of formation is given. The first problem is to write out the first 15 numbers in the Fibonacci sequence and find the ratio of each number to the previous number. The students are to use a calculator to approximate each ratio to six decimal places. Then the students are asked to compare their ratios with the value of the golden ratio.

The authors continue by writing that the golden ratio and $\sqrt{5}$ are both irrational numbers, but you can combine them in two surprising expressions to find *any* number in the Fibonacci sequence. Then formulas are stated when n is odd and when n is even. While a^n has been used for general a, it is not used here with $a = -1$ to give a formula which holds for both n odd and n even, so separate formulas have to be given. The students are asked to find the thirtieth number in the Fibonacci sequence, but not to show that the defining property of Fibonacci numbers is satisfied by the numbers stated in the two formulas. This is just one of many places where teaching students to think is passed over in favor of having them do calculations which do not lead anywhere.

A little later in this section, students are asked to use a graphing utility to repeatedly multiply the matrix A

$$A = \begin{pmatrix} 1 & 1 \\ 1 & 0 \end{pmatrix}.$$

The first part is to find A^2, A^3, A^4, A^5, and A^6, and then to describe the patterns which they see in the entries of these matrices. Leaving aside the fact that a "graphing utility" is not what was meant, the students should not use any kind of computing aid to do this. They need to do the calculations by hand to try to understand what occurs. There is more to this than just spotting the Fibonacci numbers as the entries. Students should understand how these numbers arise, how the defining relation for the Fibonacci numbers leads to their occurrence, and be able to explain this in words. It is too early to introduce formal induction arguments, but this example can be used as a way to lead up to induction in a couple of years. One of the reasons for having books appear in a series is to consider just such points.

If something like this were done often, the claim of teaching the students to think could be justified. It is not done at all.

Pseudo-Science and Bad History

The first thing one notices when looking at *Focus on Algebra* is the lavish use of color illustrations, frequently irrelevant to the mathematics being discussed. These illustrations can be very distracting, and some high school students have remarked on this to their teachers.

At the start of Chapter 3, Section 2, there is an illustration of a wooden carving from Mali. The text is about the Dogon and their interest in astronomy. Part of what is stated is relevant to the mathematics to be considered in this section, but part of it is totally inappropriate in a school text. "Anthropologists studying the Dogon in the 1940s reported that without the aid of telescopes or other instruments, the Dogon had discovered that Jupiter has satellites, and that Saturn has rings. Neither fact is apparent to the naked eye. Even more amazing, the Dogon claimed that an invisible star of enormous density orbits the star Sirius once every 50 years. Not until 1925 had astronomers discovered that a so-called 'white dwarf' — a dark, tiny, and incredibly dense star — circled Sirius." The reference in the margin of the teacher's edition is to an article in National Geographic. A fuller reference would be [10]. It is very unlikely either the students or the teacher will be able to explain what happened to Marcel Griaule, and how he was fooled by Ogotemmeli. I don't know if Ogotemmeli read French and had seen a French astronomy book, as was suggested to me by an anthropologist I asked about this claim, or if the possible reason given in the margin of the text about the Dogon learning of these facts from outsiders before contact with anthropologists is true. However, this type of pseudo-science has no place in a mathematics book. See [7] for other comments on this page.

There are some poor historical comments. The section on the golden ratio has the following: "It is said that Theano, wife of Pythagoras, did original work on the golden rectangle." I asked the mathematical historian Roger Cooke about this. Here is part of his reply:

> The 'it is said that' phrase ought to be punishable by death, likewise 'man sagt, dass' and 'on dit que'. All that is known of Pythagoras is due to later commentators, mostly Diogenes Laertius (3rd century AD, 700 years AFTER Pythagoras). It's only due to DL that we know of Pythagoras' wife and children. I don't recall offhand that he mentioned any work done by the wife, but he may have. I'm not sure he's a reliable source, since he includes two previous incarnations in his life of Pythagoras and gave two contradictory accounts of his death.

The list of reviewers of this book does not contain either mathematicians or mathematical historians. Both should be used to check the text for accuracy.

What about the Algebra?

What about some of the topics which have traditionally been a major part of a first year algebra course? Factoring is one, and a very good eighth grade teacher I have been corresponding with claims that it takes at least six weeks for students to learn this well. At the same time they are learning this, they are finally learning the arithmetic many of them have not learned well enough. Multiplication of binomials comes in Chapter 8, and it starts with algebra tiles. Here is a description, but it pales in comparison to seeing them in print. The number 1 is represented by a small yellow square. A red square of the same size represents -1. A yellow rectangle of the same width but longer represents x, and a red one of the same size represents $-x$. To represent x^2, a yellow square is used which has the same length as the long side of the rectangle which represents x. The set of tiles I bought was made carefully so that the length of the rectangle is not an integer multiple of the width, so that students will not think that x can be constructed from a fixed number of ones. These are first used to solve linear equations. There is a reference to work by Confrey and Lanier [6] that is said to show that many students at all secondary levels are still at the developmental level where learning must be facilitated by concrete and pictorial representations of concepts. I read the paper on which this claim is said to come and the summary is not accurate. They said that some students seem to benefit from this. From comments from teachers, it is clear that there are also students for whom use of algebra tiles is completely inappropriate. These objects have a very restricted use, for there are few equations which can be modeled with a moderate number of tiles. Their restricted usefulness is only one of their drawbacks. They can become a crutch which makes it hard to progress to a higher, more abstract level.

The reason given for using algebra tiles is to help students who seem not to be ready for the abstraction which comes with the use of letters to denote objects. A better solution to this problem would be to gently start using letters in elementary school, as is done in Russia. See [20] and [25]. While students who are having trouble with mathematics are being shortchanged, the recent TIMSS results suggest that our better students are equally disadvantaged in comparison to their peers in other countries. In the British TIMSS report, it is claimed that 25% of the US eighth grade students are in the bottom 25% of students internationally, while only 18% are in the top 25% [15, p. 22]. In [2, p. 31] it is claimed that only 5% of the US students in the eighth grade sample are in the top 10%, and 45% in the top half. This should be contrasted with the results from Singapore, where 45% are in the top 10%, 74% in the top 25%, 94% in the top half, and only 1% in the bottom 25%. Their texts are very interesting, and one does not see the irrelevant illustrations that clutter up our texts. Algebra tiles are not used.

When multiplication of binomials is introduced, the first example is $(x+2) \times (x+5)$. Algebra tiles are laid out on the outside of a rectangle with sides $x+5$ and $x+2$, and students are asked to fill in the rectangle using one x^2 tile and as many x-tiles and unit tiles as needed. It is here that the fact that the long and short lengths are not related by small integers is useful. One other drawback about algebra tiles is that students would tend to get the idea that x is larger than 1, and so is x^2. In fairness, the authors also use the distributive rule to do this multiplication, but only as a secondary method.

In the margin on page 598, FOIL makes its usual appearance. Almost all college students know what FOIL means, but many of their professors do not. It is an acronym used to illustrate $(a+b)(c+d) = ac + ad + bc + bd$, of First, Outer, Inner, Last. Here it is illustrated by $(2x+1)(x-3)$ rather than the usual $(a+b)(c+d)$. I have asked many people why FOIL is used, and have yet to hear a good explanation. Betty Phillips told me that she and Glenda Lappan have been trying to get rid of FOIL for over 20 years. Gail Burrill told me that FOIL was outlawed at Whitnall High School where she taught for many years. Yet, it is in a textbook she coauthored. It is in a series of video tapes on algebra which Sol Garfunkel had made. When I asked him about this, he said that he had not thought hard about it, and just used what others had used. He now thinks this was an error. At least in the present case, FOIL is not in the book the students use. However, it is presented as

Tips from Teachers: You may want to teach the students the FOIL method ... as a mnemonic for multiplying two binomials....

Factoring is then done. First, this is done with algebra tiles. Then polynomials of degree two are factored by inspection, as well as by the two special cases; using the difference of two squares and using $(a+b)^2$. On one page, it is suggested that the teacher multiply $(x-1)(x^2+x+1)$ and $(x-1)(x^3+x^2+x+1)$, and then ask the students to guess what the result will be when the next case is considered. However, there is no follow-up to this work, so what has been started will be lost for almost all students. The need to come to closure with what is started is not one of the strengths of our current reforms. The same failure to follow up occurs with respect to Pascal's triangle on the same page.

The work on factoring is usually a prelude to quadratic equations, and is so here as well. Quadratic functions and equations are the subject of the next chapter. The first method of solving quadratic equations is with a graphing calculator or with a graph on a computer screen. The possibility of drawing a graph by hand is mentioned. The word "parabola" is used, but the geometric definition is not mentioned. A parabola is just the graph of the quadratic equation

$$y = ax^2 + bx + c.$$

There are problems which ask students to explain something. For example, problem 12 on page 642 is:

The graph of $y_1 = rx^2 + 3$ is a parabola that is much wider than that of $y_2 = sx^2 + 3$. Which is larger, r or s? Explain.

The answer in the margin of the teacher's edition is "$|s| > |r|$; The larger the absolute value of the coefficient of x^2, the narrower the graph." Is it any wonder that our students in college have trouble explaining something when they are told that this statement is an explanation? The answer could have said something about why this is true, such as: "if $s > r > 0$, then it takes a smaller value of x to reach a given height for y_2 than for y_1."

Other ways of solving quadratic equations are introduced. Tables of values of the quadratic function are one way, factoring is another. Not all quadratic equations have rational roots, so a method using square roots is introduced, and used to solve equations easily reducible to $x^2 = a$. However, nothing is said about how square roots can be computed. The least that should have been done is to explain how to take square roots by bisection. A very nice treatment of square roots and cube roots from a historical perspective is contained in [11]. This is a marvelous book, written over a ten year period by a physician. It puts us to shame that a first rate book like this is written by an amateur rather than by a professional mathematician or mathematics educator.

Finally, there is a section on "The Quadratic Formula". Here is a quote from this section:

Today most people solve quadratic equations using a formula that is based on a geometric model. The *Quadratic Formula* is one of the most important formulas in mathematics.

The quadratic formula is stated, then an example is given of its use. Students are then given problems to do using it. However, there is no derivation, nor is anything said about what the "geometric model" mentioned above is. This must mean the old way the Babylonians used to solve quadratic equations. I asked one of the authors why the quadratic formula was not derived and was told the following. Quadratic equations come near the end of the usual algebra course and many teachers do not give a derivation of the quadratic formula. Since it is important that students see a derivation of it, and most states now require three years of mathematics, the derivation was put off until the next algebra book in this series *Focus on Advanced Algebra*. In Wisconsin, two years of high school mathematics are required, not three. I was told by another person associated with this series that the real reason was that the quadratic formula has to be in a first year algebra book before the book can be considered for adoption in Texas. Nothing is said in the Texas guidelines about what has to be done with the quadratic formula, so a derivation does not have to be given. The authors have moved systems of linear equations from the second algebra book to the first, and they felt that the derivation of the quadratic formula could wait. My cynical view is that systems of linear equations allow matrices to be used to solve them, and graphing calculators can be used to solve systems of two by two linear equations, including finding inverses of matrices.

Focus on Algebra is far from the worst of the new books. However, it contains a representative sample of what is being called "a mile wide and an inch deep". Another similar book is [26]. This book has much more use of algebra tiles. A review of this book is posted on the World Wide Web; see [22]. Here is one quote from this review.

> Most of the problems of the book end with the word: "Explain." But the
> book or the teacher edition never offers any explanation.

Notice the similarity between these books. They ask for explanations, but do not teach the students how to explain things.

To put the treatment of the quadratic formula into perspective, consider the following.

Math Camp is a one month program for bright high school students. At Math Camp, in August, 1996, I mentioned that some of our new programs do not give a derivation of the quadratic formula in the first year algebra course, and some do not give this until 12th grade, when many students are no longer taking mathematics. A young woman from Turkey expressed surprise, since her class had been taught this in seventh grade.

Is Another Revolution Coming?

There are many other examples which could be given. Let me close with two examples of what I hope will be in the next mathematics education changes.

First, consider trigonometry. There are just a few central ideas behind trigonometry. The first is the idea of similarity. This allows one to define the trigonometric functions so that they apply to all right triangles as functions of only one variable. Right triangles are determined by the Pythagorean theorem, so combined with similarity it is possible to define sine and cosine as the coordinates of points on the unit circle. The second important fact is decomposition. Any triangle can be decomposed into a pair of right triangles. This allows the trigonometric functions to be used in the study of arbitrary triangles, and allows triangles to be used in the study of polygons. One can use the invariance of the circle under rotations to prove the addition formula for cosine, but there are very simple direct proofs using decomposition. See the forthcoming book [8] by I. M. Gelfand and Mark Saul for some examples. Everything in elementary trigonometry is a corollary of these few facts, but not such easy corollaries that one can stop here. The usual facts need to be learned, and technical skill needs to be developed and practiced. However, all of this is much easier if the fundamentals are always kept in mind.

Finally, there is an important book [19] coming out shortly. One chapter deals with the computational problem of $1\frac{3}{4}$ divided by $\frac{1}{2}$ and attempts to illustrate this by story problems. One of the Chinese teachers said she would not use division by $\frac{1}{2}$ to illustrate division by fractions, since it is easy to see the answers without dividing by fractions. She suggested using $1\frac{3}{4}$ divided by $\frac{4}{5}$, and gave a problem exemplifying this calculation.

Contrast this story with the following passage in the NCTM Curriculum Standards, page 96. "The mastery of a small number of basic facts with common fractions (e.g. $\frac{1}{4} + \frac{1}{4} = \frac{1}{2}$; $\frac{3}{4} + \frac{1}{2} = 1\frac{1}{4}$; and $\frac{1}{2} \times \frac{1}{2} = \frac{1}{4}$) and with decimals (e.g. $0.1 + 0.1 = 0.2$ and $0.1 \times 0.1 = 0.01$) contributes to students' readiness to learn estimation and for concept development and problem solving. This proficiency in the addition, subtraction, and multiplication of fractions and mixed numbers should be limited to those with simple denominators that can be visualized concretely or pictorially and are apt to occur in a real-world setting; such computation promotes conceptual understanding of the operations. This is not to suggest, however, that valuable instruction time should be devoted to exercises like $\frac{17}{24} + \frac{5}{18}$ or $5\frac{3}{4} \times 4\frac{1}{4}$, which are much harder to visualize and unlikely to occur in real-life situations. Division of fractions should be approached conceptually. An understanding of what happens when one divides by a fractional number (less than or greater than 1) is essential."

I wrote Liping Ma to ask for comments on this quotation from the NCTM Standards. She replied as follows:

I would like to claim some interesting and important relationship between basic facts with common fractions (e.g. $\frac{1}{4} + \frac{1}{4} = \frac{1}{2}$; $\frac{3}{4} + \frac{1}{2} = 1\frac{1}{4}$; and $\frac{1}{2} \times \frac{1}{2} = \frac{1}{4}$) and with decimals that can be visualized concretely and those much harder to visualize and unlikely to occur in real-life situations. In fact, without a conceptual understanding of the former, it will be unlikely for one to understand the latter. However, unless one's understanding of the former is deepened and solidified by the latter (which is not as hard as people imagine), the primary conceptual understanding is still very limited and superficial and therefore too fragile to make connections to other concepts of the subject. So, students' mathematical power will be generated from a connection of the "basic facts" and "abstract concepts", rather than emphasizing or ignoring either of them.

Ma's book may be the start toward a balanced view of mathematics education which we have long needed. Like a stool which needs three legs to be stable, mathematics education needs three components: good problems, with many of them being multistep ones, a lot of technical skill, and then a broader view which contains the abstract nature of mathematics and proofs. One does not get all of these at once, but a good mathematics program has them as goals and makes incremental steps toward them at all levels.

References

[1] Lars Ahlfors et al, "On the mathematics curriculum of the high school", American Mathematical Monthly, **69** (1962), 189–193; Mathematics Teacher, **55** (1962), 191–195.

[2] Albert E. Beaton et al., *Mathematics achievement in the middle school years: IEA's Third International Mathematics and Science Study (TIMSS)*, TIMSS International Study Center, Boston College, Chestnut Hill, MA, 1996.

[3] California State Board of Education, *Mathematics framework for California public schools*, California Dept. of Education, Sacramento, 1992.

[4] California State Board of Education, Mathematics Program Advisory, http:// wwwgoldmine.cde.ca.gov/ci/branch/eltdiv/mathadv.htm.

[5] Randall I. Charles et al., *Addison-Wesley Secondary Math, Focus on Algebra*, Addison-Wesley, Menlo Park, CA, 1996.

[6] Jere Confrey and Perry Lanier, "Students' mathematical abilities: a focus for the improvement of teaching general mathematics", *School Science and Mathematics* **80**, 549–556.

[7] Martin Gardner, "The new New Math", *The New York Review of Books*, September 24, 1998, 9–12.

[8] I.M. Gelfand and Mark Saul, *Trigonometry*, Birkhäuser, Boston, to appear.

[9] I.M. Gelfand and A. Shen, *Algebra*, Birkhäuser, Boston, 1993.

[10] Marcel Griaule, *Conversations with Ogotemmeli*, Oxford Univ. Press, 1965.

[11] Jan Gullberg, *Mathematics: from the birth of numbers*, Norton, New York, 1997.

[12] Geoffrey Howson, *Mathematics textbooks: a comparative study of grade 8 texts*, Pacific Educational Press, Vancouver, BC, 1995.

[13] Marianne M. Jennings, "'Rain Forest' algebra course teaches everything but algebra", *Christian Science Monitor*, April 2, 1996.

[14] George Gheverghese Joseph, "The crest of the peacock", Penguin, London and New York, 1990.

[15] Wendy Keys, Sue Harris and Cres Fernandes, Third International Mathematics and Science Study, First National Report, Part 1, Achievement in Mathematics and Science at Age 13 in England, National Foundation for Educational Research, Upton Park, Slough, Berkshire SL1 2DQ, England, 1996.

[16] Morris Kline, *Why Johnny can't add: the failure of the new math*, St. Martin's Press, New York, 1973.

[17] Kunihiko Kodaira, *Japanese Grade 9 Mathematics*, 1984, translated by Hiromi Nagata, Univ. of Chicago School Mathematics Project, Chicago, 1992.

[18] Li Yen and Du Shiran, *Chinese mathematics: a concise history*, translated by John N. Crosley and Andrew W. C. Lun, Clarendon Press, Oxford, 1987.

[19] Liping Ma, *Knowing and teaching elementary mathematics: teachers' understanding of fundamental mathematics in China and the United States*, Lawrence Erlbaum Associates, Hillsdale, NJ, to appear in 1999.

[20] M. I. Moro and M. A. Bantova, *Russian Grade 2 Mathematics*, translated by Robert Silverman, Univ. of Chicago School Mathematics Project, Chicago, 1992.

[21] Calvin Moore, "California's math reforms aim to inspire economy of thought", Sacramento Bee, October 18, 1995.

[22] Motohico Mulase, review of [26]. See http://ourworld.compuserve.com/homepages/mathman/pinkbook.htm.

[23] National Council of Teachers of Mathematics, Curriculum and Evaluation Standards, Reston, VA, 1989.

[24] R. Sridharan, Review of [14], *Current Science* **70** (1996), 753–754. Reprinted in *Resonance* **1**:6, June 1996, 90–93.

[25] Zalman Usiskin, "Doing algebra in grades K-4", *Teaching Children Mathematics*, 1997, 346–356.

[26] A. Wah and H. Picciotto, *Algebra: themes, concepts, tools*, Creative Publications, Mountain View, CA, 1994.

RICHARD ASKEY
UNIVERSITY OF WISCONSIN
DEPARTMENT OF MATHEMATICS
480 LINCOLN DRIVE
MADISON, WI 53706-1313
UNITED STATES
askey@math.wisc.edu

Contemporary Issues in Mathematics Education
MSRI Publications
Volume **36**, 1999

Instructional Materials
for K–8 Mathematics Classrooms:
The California Adoption, 1997

BILL JACOB

Does 30 divide the product 36×45? If you believe it does then you should be concerned. The California State Board of Education does not appear to. In fact, they explained their reasoning in a public document and used it as justification to reject instructional materials for California students.

The year 1997 was an intense one for the California K–12 mathematics education community, with three major events. Each involved controversy, and each involved university research mathematicians. Drawing by far the most media attention — see, for example, [Co] or [La] — was the development and adoption of California's first Mathematics Standards (and Language Arts Standards). During this same period a second group met to prepare a first draft of a new "Mathematics Framework for K–12", a process which is conducted every seven years and is not yet complete. (A Curriculum Commission version will go to the State Board during 1999.) Finally, between February and September, a follow-up adoption of K–8 mathematics materials was conducted (the primary adoption occurred in 1994). The controversy surrounding this selection is the subject of the present article. The author was a member of both the 1994 and 1997 adoption panels.

California's selection of K–8 instructional materials is especially significant for a number of reasons. Unlike text selections for grades 9–12, where there is a long tradition of teacher autonomy, no state-wide timetable, and which in practice are usually made by individual schools, the K–8 selection determines the materials that most K–8 teachers use and greatly influences what is taught in schools. Further, it provides an important resource for understanding what the State Board values most in educational practice. Finally, the views expressed by the Board embody many of the beliefs at the heart of a growing national debate. US Education Secretary Riley [Ri] has expressed his concern that a failure to resolve the current "math wars" will harm the nation's students, and those seeking middle ground to resolve the conflict should find this case study informative.

Some Key California Events between 1985 and 1997

California has an extensive process for selecting instructional materials. The process includes a three-month review by the Instructional Resources Evaluation Panels (IREP) and three months further review by the Curriculum Commission (CC) which makes recommendations to the State Board of Education (SBE). All discussions and decisions are made at public meetings and must use criteria approved 30 months in advance by the SBE. The final decision rests with the SBE. A sketch of this process can be found in the appendix to this article. We set the stage for examining the 1997 adoption by reviewing a few key events.

In 1985 the SBE adopted a new Mathematics Framework [MF85] and new Instructional Materials Criteria, imprinting a different thrust to mathematics education in California. However, as noted in [BC] and [Su], the notions of the curriculum embodied in the new Framework were unfamiliar to many teachers, and apparently to the publishing industry as well. In 1986, the mathematics IREP and the CC judged that none of the submitted mathematics materials met the new criteria. (The CC subsequently initiated a year-long process working with publishers to modify 10% of the lessons so the materials could be sold in California. In order to help teachers implement the new Framework, the California Department of Education published a *Model Curriculum Guide, Kindergarten through Grade Eight* [MCG] which described "Teaching for Understanding" and included sample classroom tasks.)

In 1989, the National Council of Teachers of Mathematics issued its Curriculum and Evaluation Standards, and in 1992 the California SBE followed with a new Math Framework [MF92] and new Criteria, aligned with the NCTM standards. This framework (still in effect at this time) has been central to the recent controversy. In 1994, the mathematics IREP and the CC recommended the adoption of 9 out of 24 submitted sets of instructional materials, which the SBE approved and to which it added three non-recommended programs. This meant that, beginning with the 1995–96 academic year, for the first time in over a decade California's K–8 districts could select new mathematics materials chosen from a list of twelve that passed the full adoption process.

Tension, however, began to mount over the changing curriculum. In November 1996 the SBE selected a Framework Committee, rejecting ten of fifteen CC nominees and adding fourteen others recommended by member Janet Nicholas, who was opposed to the 1992 Framework. (Three of the rejected ten were returned the following month after the Superintendent of Public Instruction protested the SBE's action.) During 1997, the Framework Committee met between January and August, sending a draft document to the CC on a 13–9 vote, with all 8 CC nominated members in opposition. The Standards Commission approved Language Arts and Mathematics standards for K–12 in September 1997, following a year of deliberation and considering public comment. The SBE approved the Language Arts standards in November, but approved a substantially modified version of the Math Standards in December. SBE members Janet Nicholas and Robert Trigg coordinated the revision. (According to [Mi], four Stanford University mathematics professors did the bulk of the work.) The resulting document is the subject of continuing public debate; see, for example, [NCTM] or [Wu].

The Adoption Recommendations

Between February and April 1997, the mathematics IREP examined seven submitted programs using, as required by state law, the same criteria used in 1994. (The SBE had made some changes in the criteria to reflect recent legislation, but none could be utilized in the 1997 adoption because law requires that criteria be set 30 months prior to an adoption.) The mathematics criteria were divided into six interrelated sections — Mathematical Content, Program Organization and Structure, The Work Students Do, Student Diversity, Assessment, and Support for the Teacher — and an SBE-approved Evaluation Form is used to score programs in each criterion on a scale of 1 to 5. The criteria are not a check list of topics, but instead indicate standards the materials should meet.

For each criterion there are three paragraphs, written in parallel language, describing the level that needs to be attained to achieve a score of 5, 3.5, and 1, respectively. Here is an example from the "Work Students Do" section (with corresponding scores in parentheses):

(5) "Students are consistently expected to think and reason in their mathematical work..."

(3.5) "More often than not, students are expected to think and reason in their mathematical work..."

(1) "Only occasionally are students expected to think and reason in their mathematical work... more often students are expected to follow prescribed directions to achieve a predetermined answer."

In June 1997, the CC presented its report to the SBE, recommending five of the seven programs for adoption and not recommending two. The five recommended programs were [INDS] for grades 1, 2, 5, [CM] for grades 6 and 7, [MC] for grade 5, [MT] for grades K–3, and [PH] for grades 6–8; not recommended were [MC] (grades 6, 7) and [ED] (grades 1, 2, 4, 5). The CC report detailed their view of how each program was or was not aligned with the criteria and was based upon the information in the written summary prepared by the IREP.

The State Board Action, September 9, 1997

On September 2, the SBE liaisons to the CC, Kathryn Dronenburg and Bill Malkasian, in a memorandum [DM] to the SBE, recommended that all submitted programs be approved. (One "condition of adoption" was stipulated: that the pages 92–93 from the 2nd-grade unit "Does it Walk, Crawl, or Swim?" in [INDS] be removed because a discussion of students' private fears "could be personally invasive." In this lesson students list "scary things" as part of an activity to learn to collect and sort information. The memo cites "Code section 51513 which prohibits surveys of pupils' personal beliefs in specified areas without prior, written permission from parents or guardians." In 1994 the SBE was confronted with numerous objections to questions in the language arts portion of the California Learning Assessment System test, which some parents considered personally invasive. In 1997, the board was not interested in risking a repeat of this experience.)

At its September 9 meeting, the SBE approved five programs, but rejected two CC recommended programs, [INDS] and [CM]. (Though [INDS] is a K–5 program and [CM] is a 6–8 program, they are completely distinct and were separate submissions for adoption.) Both rejected programs are published by Dale Seymour and were developed by NSF-funded curriculum projects. This decision was unexpected since most people believed the Dronenburg–Malkasian recommendation would be accepted by the full board. The basis for the rejections was outlined in two memos authored by SBE member Janet Nicholas [N1; N2] that were distributed to the SBE. The decision was reached after a short discussion at the meeting, without any public input or review of the two memos. Neither the publisher nor the CC had any opportunity to respond.

State law [Ed. Code 60200 (d)] mandates that the SBE must "provide specific, written explanation of the reasons why the submitted materials were not adopted", and the two memos [N1; N2] provide that basis. The principal reasons for the rejections were

(1) inconsistencies with code section 60200.5,
(2) mathematical errors, and
(3) problems with instructional strategies employed, including the question of whether they were "research-based".

A representative sample of these objections follow.

(1) *Inconsistencies with code section* 60200.5.   First we give the precise wording of this code.

> 60200.5. Instructional materials adopted under this chapter shall, where appropriate, be designed to impress upon the minds of the pupils the principles of morality, truth, justice, patriotism, and a true comprehension of the rights, duties, and dignity of American citizenship, and to instruct them in manners and morals and the principles of a free government. The

State Board of Education shall endeavor to see that this objective is accomplished in the evaluation of instructional materials of educational content in appropriate subject areas.

To understand the SBE application of the code we quote from [N2, p. 5, paragraph 3]:

> ... The unit on fractions has the teacher tell a story about a burglar (disguised as the school janitor) and his friend who break into the school cafeteria every night to steal pizza.[1] This problem is inconsistent with the objectives defined in Sec. 60200.5.[2]

Another citation of Code 60200.5 [N2, p. 5] describes a suggested rubric score on a partner quiz sample appearing in [CM], Teacher's Edition, "How Likely is It?", Grade 6, pp. 85–88. It reads:

> Advocating grading of assessments based upon data/information not on the test is highly questionable and inconsistent with the objectives defined in 60200.5.

(2) *Mathematical Errors.*   According to [N1, p. 4] we have

> *Materials have factual errors that distract from learning and serve to confuse teachers and students alike.*

A specific example in the subsequent paragraph reads:

> Another illustration of the potential for confusion is provided by the discussion of "Rewriting Multiplication Expressions." The text indicates that 36×45 can be rewritten as an expression with three factors such as $9 \times 5 \times 36$ or $9 \times 30 \times 6$. The number 30 is *not*, however, a factor of either 36 or 45.

At the bottom of this same page we find the following example.

> On more than one occasion, the books seem to have a difficult time distinguishing between a *ratio* and a *fraction*. These are often difficult distinctions for young students and the factual error in the text makes it even more confusing. By way of example, in the unit "Writing Fractions as Decimals" the scores of ball players' free throws are presented as: Angela: 17 out of 25, Emily: 15 out of 20, e.g. — ratios not fractions.[3]

Two other similar examples detailing the ratio-fraction distinction are also listed (ibid., p. 66b, and "Comparing and Scaling", Grade 7, p. 39). Four typographical

---

[1] [CM], Teacher's Edition, "Bits and pieces II", Grade 6, pp. 46, 50.

[2] In this problem the "notorious Pizza Pirate", on successive nights, repeatedly "gobbles down half" of the pizza that the students in the story keep in the cafeteria freezer. Working on this problem, students must iteratively compute $\frac{1}{2} \times \frac{1}{2} \times \frac{1}{2} \times \cdots$.

[3] [CM], Teacher's Edition, "Bits and Pieces I", Grade 6, p. 54.

errors are noted in the Investigations' teacher materials for [INDS] (none in any student materials), each of which are readily recognized as such and is easily repaired. (Typographical errors occur in new programs, the best and worst alike. In 1994, some Silver Burdett software was so buggy that IREP members couldn't evaluate it, yet the SBE approved the program (which was not recommended by either the IREP or the CC) assuming the problems would be fixed. In 1997 the SBE approved *Mathematics in Context*, which the IREP was required to reject because it was incomplete and the pre-publication editions had numerous blank pages. In fact, in [DM] a footnote reads "Recently, we received an opinion from Deputy General Counsel Roger D. Wolferz expressing the view that the teachers' materials in the EB program should be disqualified...". Apparently the SBE was willing to assume that these pages would be filled in correctly. The board's inconsistency here is quite shocking.)

(3) *Problems with Instructional Strategies.* In its opening discussion [N1, p. 1] (and reiterated in [N2, p.1]) it is stated that "The Commission's report does not identify or mention the method it used to determine that the submissions recommended for approval actually *incorporate principles of instruction reflective of current and confirmed research* (60200 c-3)." [Emphasis added here and in the next two quotations.]

Specific examples cited include [N2, p. 3] the following discussion of Grade 1:

> The program's fundamental "theory" is that the students will learn the definitions of mathematical terms exclusively from hearing them in conversation. The Teacher's Edition states that students are not asked to learn new definitions of mathematical terms and it states that "*This approach is compatible with current theories of second language acquisition, which emphasize the use of new vocabulary in meaningful context while students are actively involved with objects, pictures and physical movement.*"[4] Current and confirmed research does not support the program's claim that the explicit teaching of definitions and terms is counterproductive to students mastering fundamental skills in mathematics. The law specifically requires instructional materials to be based on current and confirmed research rather than theories.

As a second example, [N2, p. 4] looks at Grade 6 and states:

> The program "*does not teach specific algorithms with rational numbers. Instead it helps the teacher create a supportive environment for students to grapple with interesting problems in which ideas of fractions, decimals and percents are embedded...*" I know of no current and confirmed research that supports the supposition that continuous exclusion of algorithms is beneficial to a student's mathematical knowledge and learning.

---

[4] Teacher's Edition, "Survey Questions and Secrets", Grade 1, pp. 1–20.

Discussion of the Board's Objections

We next consider the SBE objections to the Dale Seymour programs and compare them to the actual criteria for adoption. (California law requires that the IREP, the CC, and the SBE *all* use the same SBE-approved criteria, which must be published 30 months prior to an adoption.) Readers concerned more with the general issues raised by this controversy than with the details of the California math wars may wish to skip this section.

(1) *Inconsistencies with code section* 60200.5.   It is hard to understand how "principles of morality, truth, justice, patriotism..." relate to "scary things" or "pizza pirates." As I wrote in my public memo [J] to the CC, "If a six-sentence story about a 'pizza burglar', which is used to appropriately set up a good problem, is somehow unpatriotic, how is the Watergate story to be handled in history adoptions?"

So what is going on? It appears that Sec. 60200.5 is used to cover the board's objections to the use of "non-mathematical subject matter." During 1994 the SBE did have to confront complaints about "inappropriate subject matter" in its statewide testing program (the California Learning Assessment System, CLAS), and subsequently Governor Wilson vetoed funding for the tests. On September 9 the SBE did mention CLAS during its adoption discussion, and this seems to have been an important factor. California has an established procedure for handling perceived inappropriate subject matter. A Legal Compliance Committee of community members and parents studies all submitted material and reports to the California Department of Educations (CDE) and the SBE. The IREP and CC are informed that if they note something unusual it should be passed to CDE staff. During the 1997 IREP deliberations our panel did note one area of concern in the Encyclopedia Britannica program, where in the Grade 6 "Made to Measure" unit, students measure each other's body parts. We note that the Dale Seymour programs had each passed legal compliance, and that the SBE did not address the legal compliance concerns when it approved the Encyclopedia Britannica program.

Why can't the SBE discuss the real issues? For example, in any particular instance, how the use of context is useful (or not useful) to mathematical learning is a crucial consideration. In fact, this issue plays a central role in the adoption criteria—the "Evaluation Form, Mathematical Content" (paragraph 2), "Program Organization and Structure" (end of paragraph 2), and "The Work Students Do" (end of paragraph 3) each deal with appropriate use of context. Unfortunately, the SBE seems to have ignored these criteria.

In the quiz sample discussion, the SBE assertion [N2] that "Advocating grading of assessments based upon data/information not on the test..." greatly misrepresents what the program says. In the teacher's manual it is explained that, when assessing student thinking, one needs to understand how students express their ideas. During this one-paragraph discussion of the assessment sample one finds the statement, "I made this interpretation based upon the way my students talk about events in class" ([CM], "How Likely is it?", p. 85.), which presumably triggered the SBE objection. Nowhere is it suggested that the teacher assess these students' work using data/information not on the test. In fact, the program's approach is compatible with the criteria. (See [MF92, p. 183], which states "The materials include suggestions to the teacher concerning how to... Observe, listen to, and question students while they work and how to keep track of insights about the students they may have.") For the SBE to cite code 60200.5 in this context, and not refer to either the Assessment or Support for the Teacher criteria in their discussion, makes no sense. Another source of discomfort with this assessment discussion [N2, p. 5] is the high score given a response to the first question "From the description of the game, do you and you partner think...?". The students' first thought is wrong, but after starting the project they quickly correct their thinking. The issue becomes, if you ask students to describe their initial ideas, are they penalized for an error that they subsequently fix? Apparently the SBE believes they must be.

(2) *Mathematical Errors.*

In considering the question of whether 30 can divide the product 36×45, the exact beliefs of the SBE are not clear. It appears that the author of [N2] fails to understand the mathematics. Possibly some different discussion of the problem was expected, but a careful reading of the teacher's manual reveals no lack of mathematical clarity. My interpretation is that the SBE wants arithmetic skills structured according to some rigid, predetermined order, and the formulation of this question didn't align with their view of this ordering of skills. If so, this view greatly limits expectations for student understanding and is not compatible with the criteria. See "Evaluation Form, The Work Students Do", paragraphs 1, 3, and "Program Organization and Structure", paragraph 2.

The problems cited by the SBE dealing with ratios and fractions are correctly formulated and are consistent with common practice. The question posed for students in the free throw example cited above is "which player should the coach select to shoot the free throw?" This problem is assigned after students discuss, in class, the notion of "success rate". Observe that the statement of the problem does not call the information being presented a "fraction" — it merely lists the events that occurred. Nowhere is any claim made that these ratios are fractions and interpreting this information correctly in context is the point of the problem.

In fact, the treatment of the concept of proportion in [CM] teaches students correct usage of units in applications, and later connects this skill with numerical, graphical, and symbolic representations of linear functions by the end of Grade 7. The author found the mathematical treatment in full alignment with the criteria, and also found the SBE claims to be unsubstantiated. (Writing about the confusion some educators have about ratios and fractions, U.C. Berkeley Mathematics Professor H. Wu in [Wu] says "Some educators, it is said, have begun to advocate that fractions are not ratios. If so, then we must redouble our efforts not to allow such ideas to creep into any mathematics standards.") *It is a major concern that the SBE member who, in 1997, arguably has wielded more power over California's mathematics education future than anyone else would author a public document leading us to seriously question her understanding of middle school mathematics.*

(3) *Problems with Instructional Strategies.*   To be sure, the issue of instructional strategies is a hot topic these days. But we must first realize that, regardless of the SBE's current beliefs about the "best instructional strategies", the 1997 adoption was, by law, required to follow the criteria set in the 1992 Framework. This Framework claims that its instructional strategies are research-based; see, for example, [MF92, pp. 32–33]. So, although the Commission's report may not explicitly identify its method for complying with 60200 c-3, the answer is built into the adoption process and is implicit in the state board approved Framework and Instructional Materials Criteria.

The criticism in [N2] about learning of vocabulary in first grade greatly distorts what is actually said, to wit, "mathematical vocabulary is introduced naturally during the activities." [INDS], "Survey Questions and Secret Rules", pp. 1-20. In this program teachers do teach proper mathematical vocabulary quite explicitly, but do so in settings where the students will use the terms immediately. In addition, the unit contains a section supporting teachers who work with students with limited English proficiency; it gives references to support its approach (ibid, p. 108).

The statement about "continuous exclusion of algorithms" in Grade 6 [N2, p. 4] misrepresents the Connected Mathematics Program. This program explicitly teaches the importance of developing facility with algorithms for computation. The first words of the sentence quoted are "This unit does not...", and for the memo to state "The program does not teach specific algorithms..." is deceptive. The unit cited is the first of *two* units on fractions, and the arithmetic operations are studied in detail in the second. There one finds, "Talk to your class about what an algorithm is in mathematics" ([CM], "Bits and Pieces II", p. 53, "Launch"). The overview of this second unit explicitly states "We expect students knowing algorithms for computation that they understand and can use with facility" (ibid., p. 1b). The assertion of [N2] that "the program's presumption seems to be that mathematical 'meaning' is inconsistent with the acknowledgment of specific algorithms" completely lacks justification and is irresponsible.

There are more examples in the memos [N1; N2] than can be given here. But the above samples are representative. As shown in detail in [J], outside of four typographical errors, the objections raised by the SBE either misrepresent the criteria or program content, or (in several instances) are based upon fallacious mathematical reasoning. *The documents (all of public record) show that the SBE*

*failed to be even-handed in its review of mathematics materials, and was willing
to compromise its credibility in order to take sides in California's "math wars".*

Issues of General Concern Raised
by the Adoptions Controversy

Although the Standards debate in California captured media attention, the
SBE's actions on adoptions may be the best indicator of its understanding and
beliefs about mathematics education. I hope mathematicians and educators
will look closely at the views expressed by California's SBE in their adoption
documents as they think about the welfare of K–8 students.

What else can we learn from the California's 1997 adoption? Here are some
further thoughts.

**The distinction between the intended curriculum and the received cur-
riculum.** The California Mathematics Instructional Resources Evaluation Form
stresses mathematical content throughout, but asks evaluators to view materials through several lenses, such as "The Work Students Do" (a high priority)
and "Support for the Teacher", as well as having a criterion actually labeled
"Mathematical Content". This is an approach not so common in the university
mathematics culture. The criteria provide distinct, yet highly interrelated, ways
to view the mathematics children will think about in school. One message this
sends is that even the most brilliant content exposition may or may not lead
students to mathematical understanding. K–12 educators have known this for a
time. Yet many mathematicians writing about K–12 devote heavy emphasis to
textbook exposition of theory. Isn't this approach shortsighted? Consider the
conclusion of one analysis [FP, p. 418]: "the results of this investigation challenge
the popular notion that the content of math instruction in a given elementary
classroom is essentially equal to the textbook being used." In [BC, p. 251], one
finds "We certainly saw many different versions of 'following the textbook' in
the California classrooms that we observed."

Shouldn't we be equally concerned with the work students do (e.g., is it rote,
or is it thoughtful and utilizing understanding?), in addition to how texts lay out
the theorems? Shouldn't we begin to think about what educators call *pedagogical
content knowledge* (see [Sh] or [K] for extended discussion), which differs from
general knowledge of pedagogy and content knowledge and instead refers to the
ability to represent ideas in a way that makes them understandable to students?
I think we mathematicians need to find a better balance, both in our discourse
about university courses as well as in our interactions with the K–12 community.
Only by viewing ourselves as members of a team (not arrogantly as the arbiters
of true mathematical content) can we play a valuable role.

Using non-mathematical contexts to teach mathematical ideas. The
September 9 actions of the California SBE have shown that the lines between

the issues of factual accuracy, principles of pedagogy, and appropriate non-mathematical subject matter have at best become blurred, and at worst disastrously confused. As long as mathematics curricula use non-mathematical contexts (whether in word problems or extended investigations), this difficulty will not disappear. Mathematicians can help by clarifying the important, big ideas of mathematics. But if this results in an emphasis on formalism to the exclusion of thinking about how all students (not just the top few) can access ideas, it can cause great harm. We need to recognize that, without a trained eye, our adult perspectives may tell us that some activities seem to trivialize the mathematics, yet from a developmental view are an essential experience for children. (See [T] for a discussion of his early understanding of division with remainder.) While I do not understand the origin of the SBE obsession with the ratio-fraction distinction, I do know that it contributed to the rejection of a program which had as its greatest strength the development of proportional reasoning. Because of a bizarre notion of divisibility, the SBE overlooked the value of a program that expects students to engage in thoughtful problem solving—something that mathematicians value most highly. These outcomes were a tremendous tragedy. In spite of their lack of accuracy, the two documents [N1; N2] are being used in attempts to prevent districts from using these Dale Seymour programs. Standards Commissioner Bill Evers presented them both to the Palo Alto School Board prior to its March 31 meeting when they considered adoption of [INDS] in grades 1 and 2.

The importance of an open process in setting public policy. Both of California's 1997 Adoption and Standards controversies included substantial last-minute changes initiated by the SBE. In each case, a single (and the same) board member played the determining role. While the SBE has the authority to make these decisions, the haste and lack of public input into its actions has undermined public confidence in the process. (For example, Profs. Farrand and Moore state in [FM] "Our input has so little value that we have six hours to respond to" the Standards. Charles Weis, Ventura County Superintendent of Schools, writes in [We]: "In my opinion, this back-room rushed revision process, though legal, is not in keeping with the public nature of governance afforded our public education system". Luther Williams, Assistant Director for Education and Human Resources at the NSF is quoted in [La] as saying: "The board action is, charitably, short-sighted and detrimental to the long-term mathematical literacy of children in California." Superintendent of Public Instruction Eastin is quoted in [Wa] as saying, "I urge you to ignore the board's standards and reach for the higher standards.")

In the case of the Standards, mathematicians played key roles in the back-room changes. (We may be headed for a repeat performance. Janet Nicholas has recently informed the CC that a group of mathematicians will prepare the instructional examples for the Framework revision, but did not reveal who is doing the work or when it will occur. As in the case of the Standards revisions, this appears to in conflict with Sec. 11121.8 of California's Bagley–Keane Open Meeting Act, and the approach will certainly heighten the current tensions.)

In the case of the adoption process described in this article, the SBE's documents are so seriously flawed that we are led to question the integrity of the board. (In contrast, [Ro] describes a potentially analogous process in New Jersey, where the time and care invested paid off, resulting in an apparent consensus absent in the California situation.) We need to demand that state decisions be open and public, that time and care be allocated and that the approaches taken balance the views of all parties. Without this intrinsic fairness, Secretary Riley's call for peace will be in vain.

Appendix: The California Math Wars

The main institutions and documents. Like most states, California has a *State Board of Education* (SBE) which is appointed by the Governor subject to Senate approval. The *Superintendent of Public Instruction* (SPI) is elected by popular vote and heads the *California Department of Education* (CDE). The CDE provides administrative support for the other agencies listed here (including the SBE), disseminates documents, answers questions, etc. Due to recent court decisions, the SPI has little authority over Standards, Frameworks or adoptions, with the final say in these areas now resting with the SBE.[5] Established by legislation in 1995, the *Standards Commission* has 21 members (twelve appointed by the Governor, six by the SPI, two by the legislature, plus the SPI), and prepares grade-level-specific content standards for SBE approval. Although the standards are voluntary for school districts, the state's new testing program must be aligned with the standards in 1999.

The SBE appoints the *Curriculum Commission* (CC) whose task is to oversee the drafting of Frameworks, the instructional materials adoption process, and to make recommendations to the SBE. The majority of the Curriculum Commissioners are K–12 teachers, administrators, or resource specialists, with a few from higher education. The CC nominates, subject to SBE approval, the members for two types of panels to help them with their work, the *Instructional Resources Evaluation Panels* (IREP) and the *Framework Committees*. Most IREP members are K–12 teachers or resource specialists but they also include higher education specialists. The IREP work will be detailed below. The Framework Committees spend 6 to 8 months revising or drafting curriculum frameworks, and their

---

[5]The legislative report [LHC, cover letter] called for such changes, stating that "the Superintendent of Public Instruction has assumed the role of policy maker and the State's schools are without the benefits associated with an educational policy governed by a strong state board" and then recommends "The Attorney General should file an action ...". This debate hasn't ended — see for example [Wa], which reports on the SPI criticism of the Mathematics Standards and states "(SBE President) Larsen says the board may ask (Governor) Wilson to cut (SPI) Eastin's budget as punishment."

drafts are revised, following public input by the CC prior to submission to the SBE for final approval. The SBE also approves *Instructional Materials Criteria* (the "Criteria") and an accompanying *Instructional Resources Evaluation Form* which must be used by the IREP, the CC, and the SBE in materials adoption decisions, and which by law must be set in place 30 months prior to an adoption.

The instructional materials adoption process. California's K–8 instructional materials adoptions are highly formalized, due in part to public scrutiny over the years[6]. The process is set by law [Ed Code 60200] and is sketched next in a series of nine steps.

Step 1. Adoption criteria and evaluation forms are approved by the SBE in accordance with Ed. Code 60200 and are supposed to be fleshed out in the Frameworks. By law, this approval must precede the adoption by 30 months.

Step 2. Publishers submit instructional materials to the CDE. Instructional materials include everything they expect the state to pay for, including teacher's manuals, texts, manipulatives, software, workbooks, etc.

Step 3. The SBE-approved IREP members meet for one week with a CC subcommittee for training on criteria and legal requirements. The sections of the Framework which explain the meaning of the criteria are a central focus of discussion. IREP legal instructions are to judge programs against the criteria, not to simply "pick what you might personally like." The review is not a competition. Potentially all programs can pass, or possibly no programs will pass. The week concludes with publisher presentations on their materials. All sessions are public.

Step 4. IREP members spend 8 to 10 weeks at home reviewing submissions and preparing citations to document their evaluation according to the criteria. Law prohibits any contact between publishers and panel members. (The author devoted, on the average, 20 to 30 hours preparation for each program, evaluating 8 programs between April 23 and June 18 in 1994.)

Step 5. IREP members meet for one week to deliberate and score each program on each criterion using the *Instructional Resources Evaluation Form*. Publishers are given an opportunity to answer questions, and IREP members prepare a *written* report for the CC. The meetings and the IREP report are public.

Step 6. The CC prepares its recommendations for the SBE, building upon what they learned from the IREP. Publishers may meet with commissioners during this process. This process takes approximately 3 months and includes formal public hearings.

Step 7. Concurrent with IREP and CC deliberations, the CDE prepares and notifies publishers about legal compliance issues, to which they can respond.

---

[6]See, for example, commentary by Richard Feynman in [F].

Step 8. The SBE makes its decision after reviewing the CC presentation and formal public hearings, using the criteria set 30 months in advance as prescribed in Code 60200.

Step 9. Once the SBE approves programs, successful publishers market their products and local districts generally test them through pilot programs. The local decision process can vary from several months to several years. No less than 70% of a district's Instructional Materials Fund must be spent on adopted materials, while 30% may be spent on non adopted materials. Waivers to the 70% rule can be obtained from the SBE. (Between 1994 and 1997 most mathematics waiver requests were granted.)

Discussions of the six sections of the Instructional Materials Criteria, as used in the 1994 and 1997 adoptions, can be found in Appendix A of the Framework [MF 92]. The Instructional Resources Evaluation Form (not found in the Framework) consists of a rubric for each section, which contains 3 columns characterizing top (5 points), passing (3.5 points), and bottom (1 point) scores. IREP members must come to consensus on an integral score between 1 and 5, and these scores are multiplied by a weighting and summed to produce a score out of 100. A score of 70 is considered passing. The sections and their respective weightings are:

1. *Mathematical Content*   score $\times 4$
2. *Program Organization and Structure*   score $\times 3$
3. *The Work Students Do*   score $\times 5$
4. *Student Diversity*   score $\times 2$
5. *Assessment*   score $\times 2$
6. *Support for the Teacher*   score $\times 4$

BILL JACOB
UNIVERSITY OF CALIFORNIA SANTA BARBARA
DEPARTMENT OF MATHEMATICS
SANTA BARBARA, CA 93106-0001
UNITED STATES
 jacob@math.ucsb.edu

Contemporary Issues in Mathematics Education
MSRI Publications
Volume **36**, 1999

Beyond the Math Wars

JUDITH ROITMAN

Introduction

Mathematics education is in ferment. Perhaps it always is, but disagreements have become louder; these quarrels have not only captured the general public's attention, but have been exacerbated by articles and opinion columns in the mass media. Almost nothing is considered beyond question, and many people have decided that they are in a "war" in which they need to take sides.

Wars are notable largely for the destruction they wreak. Occasionally they are unavoidable, but there are usually better ways to decide things. This particular war is a war only in the minds of those who wage it, and the supposed sides have more in common with each other than they think. Furthermore, many of the charges made against, or the descriptions of, one supposed side or the other are false.

The purpose of this paper is to puncture some of the myths lying behind the notion that there is a war going on, and to point to areas of broad agreement. Sometimes the agreement is only that a particular issue is an important one, but that is a place to start. Usually the agreement is much deeper.

At the suggestion of the editors, a glossary is provided at the end to describe some of the more technical terms in education. Misunderstanding of what these terms mean is one reason for the notion that war is necessary. Terms in the glossary are in boldface in their first appearance in the text.

This is an expansion of a talk given at the Joint Mathematics Meetings in Baltimore in January, 1998, at a special session co-sponsored by MER, AMS, and MAA. Given the rapid changes since the MSRI meeting in K-12 mathematics education, it seemed preferable to write up this talk rather than the one given over a year earlier at MSRI.

Things are Not What They Seem

The war is generally described as between reform methods and curricula —
usually based on the *Standards*[1] published by the National Council for Teachers of Mathematics (NCTM), or somewhat aligned state frameworks and standards — and traditional methods and curricula. But these descriptions are gross
oversimplifications. Let me give three examples.

1. Critics of **contextualism** in the schools conflate it with *Standards*-based
   reform, but the majority of examples in the NCTM *Standards* are not what
   contextualism recommends, even when they are superficially about something
   concrete. For example: "I have six coins worth 42 cents ..." No problem
   beginning with this phrase can be considered contextual in any serious sense.
2. **Saxon**'s books, touted as exemplars of traditional education, are also influenced by the new math movement of the 1960's and 1970's, especially with
   regard to the emphasis on fairly abstract logical and set-theoretic notions.
3. The technique of scripted **direct instruction**, often touted by opponents
   of *Standards*-based reform, is decidedly non-traditional. Even more confus-
   ingly, of the mathematicians usually identified with reform, a surprisingly
   large number (I am one of them) became interested in K–12 mathematics education through involvement in **Project SEED** or one of its offspring, which
   used techniques very close to scripted direct instruction.

Just to show you how messy things are, here are some quotes, taken both off
the Web and from published documents. See if you can identify which were
uttered by someone associated with reform, and which by someone considered
to be opposed to reform.

(i) "Elementary school students should develop rapid facility with addition and
    multiplication problems."
(ii) "What do students need to memorize? How can that be **facilitated**?"
(iii) "As I looked at the graph, I thought 'Where is what kids are learning?
    Where is the mathematics?' Nowhere in the display or in the description of
    the measure of success was there any mention of the mathematics students
    were studying and what they were learning. This worried me — it worried me
    a lot.

    ... The bottom line is that to measure whether a situation has improved,
    you have to 'show me the mathematics' (to paraphrase the popular saying) —
    look for the mathematics being taught and learned in the classrooms being
    observed. The measuring stick should focus ... on what students ... are learn-
    ing."

---

[1]There are three volumes, the 1989 *Curriculum and Evaluation Standards*, the 1991 *Professional Standards*, and the 1995 *Assessment Standards*.

(iv) "The duplication of content from year to year in mathematics texts for grades six, seven, and eight is so great that, lacking labels, it is actually difficult to arrange them in the intended order."

(v) "Whatever the reason, proof in the applications, problem-centered domain of secondary school mathematics is postponed — suppressed — downgraded ... There is even strong support for the idea that we should not presume to do much with proof at the secondary level."

(vi) "...all students can ...deduce properties of, and relationships between, figures from given assumptions."

(vii) "Differences in learning rates must be recognized and provided for."

(viii) "Problems are the life blood of mathematics."

(ix) "...the pressure now exerted by students, administrators and parents to grade on the curve, lower standards, and inflate grades."

(x) And, finally, we have a flowchart:

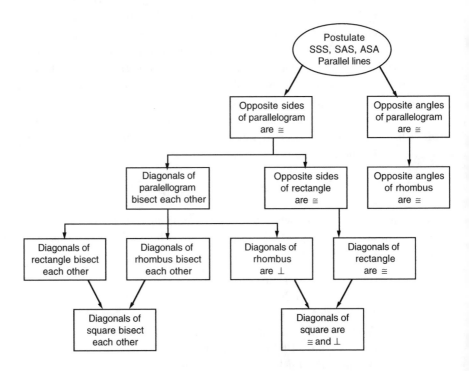

The first three quotes emphasize rapid facility with arithmetic, memorization, and the need for mathematical substance. Surely they are from one of the documents supporting traditional education. But no, the first two are from the *Professional Standards*, and the third is from an essay by Gail Burrill, President of NCTM, in the NCTM News Bulletin.

Quote 4 and 5 are from a document attacking reform written by Frank Allen,[2] a past president of NCTM who is horrified at the direction NCTM has taken. But it is the traditional texts, not the NSF-funded reform projects, that duplicate content with such gusto, and as for the charge about proof, note that quote 6, which wants all students to give proofs, is from the *Curriculum and Evaluation Standards*.

Quote 7 is problematic. Is it supporting **tracking**? Or **heterogeneous classrooms** that place different demands on different children at the same time in the same place? Since it is from Frank Allen, it is the former, but note how easily the same words could be used to support the latter.

Quote 8 could be contextualism, it could be anything, but it is Frank Allen again, as is quote 9, which charges reform with yet another thing it has nothing to do with. Whatever the origins of grade inflation, *Standards*-based reform is not one of them.

As for the flowchart, it looks like a fairly traditional (except possibly for the format) description of how theorems flow from the postulates of Euclidean geometry, and so it is, embedded in the *Curriculum and Evaluation Standards*.

So things are not what they seem. Out of context it is not easy to figure out who is asking what. Without certain rhetorical phrases that are dead give-aways (the second ellipsis in the third quote leaves out the phrase "all children", generally associated with reform; note that the association itself is a calumny against the traditionalists) it is hard to figure out who is speaking. But since much of the context is political, context can obscure what is being said. You often have to leave the context out to try to figure out what actually is being suggested for the classroom.

Politics Intrude

When politics intrude, what actually is being suggested for the classroom can get lost in the rhetoric. When the politics gets heavy both sides use the same supposed argument. Here is how it goes:

Step 1. Here are my [our] credentials.
Step 2. What you have heard so far is biased.
Step 3. Test results are actually different from what they tell you.
Step 4. What I suggest to you works, and parents know it.
Step 5. Don't be fooled by political lobbying.
Step 6. We must establish world class standards now.

Those familiar with the situation in California, where the state mathematics framework and standards (two different documents, created by nearly independent processes) elicited particularly intense political maneuvering, will recognize

---

[2] This document was signed by many critics of reform, including mathematicians, and presented at the 1996 NCTM annual meeting as an alternative to the *Standards*.

this list as a description of public testimony given by parents, mathematicians, teachers, and others, before the various boards, commissions, committees, etc., involved in the Byzantine process of shaping the curriculum of California schools. Two things are worthy of note: Steps 1 through 5 are not an argument, and both "sides" used it.

The California situation was most unfortunate in the heavy way it was politicized. Testimony about the framework and standards could take no other form than that described above, since in order to be effective it had to be political (state structure gives final power to the state Board of Education, a politically appointed body with great powers that it does not hesitate to use). The politics in California were more in the public consciousness than in most states; but, thanks to organizations like **HOLD** and **Mathematically Correct**, and the columnist **Debra Saunders**, other states have not been immune to more quiet but equally pervasive politicization of state standards. Comment on state standards in many states tends to be taken out of context and seized on by antagonists in this mythical war, making serious discussion quite difficult. Litmus tests, such as the use of calculators, are rigidly applied. Certain standards (e.g., California's and Virginia's) become battle-cries uttered by people who have no idea what is in them.

The sorts of caricatures that accompany politicization do not lead to helpful solutions to a very serious and long-standing problem, which is that by many measures American schools do not work; in particular, they do not work in math and science. This problem long predates reform. We look lousy in **TIMSS**, but we also looked lousy in **SIMSS** and in **FIMSS**.

Caricatures

These caricatures belie important underlying agreements within the mathematics community.

At the Baltimore panel on which this article is based, Frank Wang, who heads Saxon Publishers, brought up, for his horror-show example, an example very similar to what I was going to bring up. The gist of both of our examples was the same: it isn't that American kids can't calculate (in international comparisons they do well on calculations), it's that they have trouble figuring out what calculations to use if they are not told, and they don't know what the results of the calculations mean. There are two classic examples of this. "Is 10% of 81 $<$, $=$, or $>$ 9?" (American kids often don't know how to begin, even though they do fine when you ask them "What is 10% of 81?") "If you have 147 kids going on a field trip, and at most 36 kids can sit on a school bus, how many buses do you need?" (4 remainder 3, or $4\frac{3}{36}$ are typical answers.)

For another example, within the broad mathematical and mathematics education community, there is agreement that we have largely failed to help teachers learn the mathematics they need in pre-service, and we have largely failed to provide them opportunities to deepen that knowledge throughout their careers. While there is, as there should be, a good deal of debate about what future teachers should learn and how they should learn it, that there needs to be a sharp increase in teacher's mathematical knowledge is denied by no-one, and the many interesting experiments and proposals that exist cannot easily be categorized by the terms used to describe protagonists in the mythical math wars.

Even when it comes down to specifics, there is a surprising amount of agreement — everyone loves the way the Japanese teach mathematics and wants to claim them for their own; the Singapore framework and the Russian problem books have a broad range of admirers.

There are serious issues on which people differ, but how they differ on those issues does not arrange itself neatly.

What are the caricatures that keep us from seeing the problem and the areas of agreement?

Here are some of the false charges made against the *Standards*:

(i) No conventional algorithms are to be taught.

(ii) **Constructivism** rules: children must invent all of school mathematics.

(iii) Individual work is discouraged: children must not only invent all of school mathematics, but they must invent it solely by working in small groups.

(iv) The concept of "proof" is essentially eliminated.

(v) Facility with arithmetic and algebraic manipulations is discouraged.

(vi) Mistakes go uncorrected — everything is okay in **"fuzzy math."**

(vii) Contextualism rules: all of school mathematics must be motivated by real-world problems.

(viii) Teachers never lecture, they facilitate.

Here are some of the false charges made against critics of the *Standards* (i.e., traditionalists):

(i) Only conventional algorithms are allowed.

(ii) Children must do as they are told.

(iii) Only individual work is allowed — worksheet after worksheet after worksheet, with no chance for class discussion.

(iv) "Proof" means dry and dull two-column proofs in geometry of obvious statements from other obvious statements; no other reasoning is encouraged.

(v) Except for Euclidean geometry (see #4), only facility with arithmetic and algebraic manipulations is the focus of the curriculum.

(vi) Mistakes are to be corrected immediately by the teacher, without discussion.

(vii) The only word problems allowed are those for which the teacher can present a precise algorithm for solution.

(viii) Teachers only talk at students; teachers never listen to students and are insensitive to student needs.

These statements have just enough grounding in truth (it doesn't take very much) to be believable: for example, children in traditional classrooms often spend a lot of time working on worksheets, and children learning from NSF-sponsored curricula often spend a significant amount of time working in small groups. But they are also caricatures: good teachers of all sorts involve their classes in discussion, and children learning from NSF-sponsored curricula also spend a significant amount of time working on their own.

When discussion focuses on these caricatures, we become sidetracked. They are all false, and that is all that needs to be said about them. Instead of being distracted, we should get down to business.

Why Can't We Get Down to Business?

Roger Howe, a professor of mathematics at Yale University, and former chair of the AMS **ARG** (see glossary), has listed key issues in mathematics education in [2]. Here is his list:

(i) relative performance (in international comparisons)
(ii) equity
(iii) technology
(iv) demography
(v) subject matter
(vi) pedagogy
(vii) teacher preparation and certification
(viii) assessment
(ix) high performers
(x) new curricula

This is the business we should be getting down to, in a civil and professional manner. But there are at least four delusions that make it difficult to discuss these issues.

The delusion of assessment. This is the notion that there is a way to find out which pedagogical method or curriculum works.

Ron Ferguson, a professor of in the Department of Mathematics and Computer Science at Texas A&M, clearly delineated the problems of designing a fail-safe study in a message sent to the math-teach e-mail list (run by Gene Klotz' estimable Math Forum, an excellent source for references in mathematics education; see http://forum.swarthmore.edu/). He considered a simple situation: design a study of calculator use in second grade, taking into account interaction with teaching strategies. What school district wouldn't love such a study?

Here are the pitfalls Ron pointed out: First you must randomly select teachers to take part — but some won't do it. Then you have to randomly assign both calculator/non-calculator use and teaching method — but some teachers will refuse to teach by the assigned method, or refuse/insist on using calculators;

to do otherwise would go against their basic beliefs about teaching. Then you have to carefully coach the method. Then you have to carefully monitor what goes on in the classroom—no deviation can be permitted. Only after all this can you collect data (and my question is: which data do you collect?). Finally you do the appropriate statistical analysis. And even then—so what? How does what happens in second grade affect what happens in ninth grade? We may theorize, but do we actually know? And even if we did—so what? The results are statistical, they don't tell us what will work with this kid right now. And even if we knew that, we would be left with a philosophical question: do students need to be able to do addition without the aid of a calculator?

To quote Ron directly: "There is nothing quite so violent as a war based on differences in faith ... Good teaching ...is a day by day experiment in which the teacher tries to find a combination of new and old that works with some student."

This is not to say that there is nothing to be learned from research in mathematics education. There are a great many pieces of wisdom to learn. But what is "best" is not one of them.

The delusion that curriculum can be judged on the page. We need serious discussion of curriculum—from mathematicians, from teachers, from people who do research in mathematics education. The ARG reports have given us a good start on issues such as proof/reasoning/logic; algorithm; algebra and precursors to algebra; mathematical modeling/word problems; statistics. The ARG discussions have been refreshingly free of ideology: one does not argue for the importance in school mathematics of the geometric series or transformational geometry or the study of algorithms *qua* algorithms because one is a traditionalist or a *Standards*-based reformer. But within the mathematics community there is a tendency to judge on the page and not in the classroom. I have seen materials that I thought were terrible work well in a classroom; and we all have the experience of preparing mathematically elegant material only to discover that our students didn't understand one bit of it. Our judgments should be tempered by the knowledge that what is on the page is only a small part of what happens in the classroom.

The delusion that, without a lot of observation in a variety of classrooms, we still know what goes on in the schools. We don't. What happened to us n years ago or what is happening to our child right now is only a small section of a highly heterogeneous solid.

The delusion that if we want to fix it they will let us, and even pay for it. Mathematicians are not particularly welcome at many of the relevant tables, being seen variously as superfluous or arrogant. There are signs of improvement (for example, the ARGs, or the fact that six of the 24 members of the NCTM *Standards* revision writing team have done significant mathematical research), but the general statement remains.

At the MSRI meeting which is the occasion for this volume I presented my own home-made unscientific chart of who has power in mathematics education. State legislators and other politically related sorts (e.g., state school boards) were at the top; teachers were near the bottom, and mathematicians were at the very bottom.

Even within the universities resources are lacking for the mathematics education of teachers, the one aspect of K–12 education for which everyone agrees mathematicians bear some reponsibility. There is general agreement that what we do now is inadequate, and even the beginnings of an outline of what teachers should know. Hung-Hsi Wu, a professor of mathematics at Berkeley, most conveniently has an article in this volume on this issue [5] which provides one place to start, and one of the few good things to have come out of the California situation is his involvement in a relatively well-funded attempt to work with teachers. (He has an interesting preprint describing the mathematics he observed in existing teacher enhancement projects [6].) Al Cuoco, Paul Goldenberg and their colleagues at **EDC** have produced material which provides another, not incompatible, place to start (e.g., [1]), and their EDC colleague Deborah Schifter has produced interesting volumes on teacher enhancement [3; 4]. (These references are by no means comprehensive, but simply a place to begin.) The U.S. Department of Education is funding an MAA/CBMS grant to examine the issue of teacher preparation. Discussion is going on through e-mail lists, within departments, at meetings. But what is missing is institutional will. Rather than being considered a major part of the educational mission, in many universities teacher preparation is relegated far below engineering education, and teacher enhancement isn't part of the official mission at all. One of the major tasks we face is convincing our colleagues, our deans, our local school districts, and our state departments of education to change this situation.

A Cautionary Tale

Recently, on an e-mail list, there was a very long and at times vituperative debate about two versions of a problem about coins. Version 1 is "I have a nickel, a dime, and a penny. How much money do I have?" Version 2 is "I have three coins. How much money could I have?"

One group essentially said: Version 2 is terrible! What sort of answer will a teacher expect? How can a kid learning how to add 1's, 5's, 10's, and 25's be expected to solve it? They will find one or maybe two possibilities and then think they're done!

Another group essentially said: Version 2 is great! It encourages kids to list possibilities carefully! It lets kids practice their arithmetic skills! It encourages them to check other kids' answers!

Ten years ago someone might have said Version 1 is no good, but we've made enough progress so that at least that didn't happen.

Now what was wrong with this debate is that several weeks went by before it became clear what the debate was really about. The issue wasn't whether either of these problems are good or bad (they both are good.) The issue is: do you teach basic arithmetic first and then other mathematical skills later? Or do you introduce skills that will be important later, like listing things systematically, *while* kids are learning basic arithmetic? And if the latter, what is appropriate when? But until the issue was clear, we were talking past each other.

The issue was not clear because version 2 is clearly a reform-type problem. A more specific version (where we know the coins are among pennies, nickels and dimes) shows up in the *Curriculum and Evaluation Standards*, and the contrast between the two versions was actually suggested by NCTM leadership to the media to show the difference between traditional and reform problems. Version 2 thus carried a political burden — attitudes about reform became projected onto it, and it was very difficult to clearly see what we were talking about. I do not exempt myself from the influence of politics; my own early comments on it were unduly sunny (I thought first graders could handle four possible coins).

Summary

Let me sum up this paper in four sentences: There is no math war. Politicization distorts things. We have too much to work on to spend time sniping at each other. We need to get down to business.

Glossary

ARG: Association Resource Group. These are committees set up by various organizations, including many of the mathematical organizations, to assist the NCTM in revising the *Standards*. For information on the various ARG reports, see the relevant society journals and Web pages, as well as the NCTM Web page.

Constructivism: 1. The belief that knowledge is necessarily constructed, not passively received. 2. Often misinterpreted to mean that children should re-invent all of mathematics. 3. Often misconstrued (under both the rubrics of reform and of anti-reform) as a pedagogical method rather than an epistemological belief. This comes from the fact that constructivism does have pedagogical consequences. These consequences are not, as is commonly believed, a rejection of all lecturing in favor of children working in small groups, but rather come from paying careful attention to the question "How can I present this material so my students can make sense of it?" The phrase "make sense" (emphasis on "make") instead of "learn" is what makes this a constructivist question. 4. Sometimes confused with social constructivism (the belief that knowledge is socially constructed).

Contextualism: The belief that all of the mathematics we teach our students should be relevant to their lives.

Direct instruction: A pedagogical method similar to group discovery (see Project SEED), with the addition of periods of rote recitation (e.g., of arithmetic facts).

EDC: Education Development Corporation, an education think tank spun off from MIT.

Facilitate: 1. The sort of word a constructivist teacher might use. 2. Sometimes misconstrued as an invitation for the teacher to provide no direction.

FIMSS: First International Mathematics and Science Study. In the U.S., run by the U.S. Department of Education and the National Center for Education Statistics.

Fuzzy math: One of the pejorative terms (another is "New New Math") used by opponents of *Standards*-based reform to describe *Standards*-based reform.

Heterogeneous classroom: A classroom with children of varying abilities and skills.

Hirsch: 1. E. D. Hirsch, A college English professor, author of *The schools we need and why we don't have them.* 2. A list of precise curriculum goals, subject by subject and grade by grade, associated with Hirsch's criticism of our schools.

HOLD: An anti-reform parent group in Palo Alto. See http://www.rahul.net/dehnbase/hold/.

Homogeneous classroom: A classroom in which children have a narrow range of abilities and skills.

Mathematically Correct: An anti-reform parent group based in Southern California, with members all over the country. See http://ourworld.compuserve.com/homepages/mathman/.

Project SEED: A program founded in the 1970's that brought group discovery learning to elementary children in underprivileged neighborhoods. Group discovery as envisioned by SEED's founder, Bill Johntz, was scripted. The teacher (generally not a regular teacher, often a mathematics or education graduate student) prepared a sequence of questions, the answer to each being short and fairly obvious, but the entire sequence designed to lead children through fairly advanced topics, e.g., negative numbers in second grade.

Saunders, Debra: A columnist in California who generally does not like what is going on in the schools.

Saxon: 1. John Saxon, a retired army officer who wrote a series of mathematics texts and founded a publishing company to produce and sell them. 2. The Saxon textbook series. The main principles adopted by Saxon are: learning takes place in small increments; problems from earlier sections should occur in later sections; students need a lot of problems to practice on; explanation should be kept short; problems should be close to template problems.

SIMSS: Second International Mathematics and Science Study. In the U.S., it was run by the U.S. Department of Education and the National Center for Education Statistics.

TIMSS: Third International Mathematics and Science Study. In the U.S., it was run by the U.S. Department of Education and the National Center for Education Statistics.

Tracking: placing children in homogeneous classrooms.

References

[1] A. Cuoco, J. Goldenberg, J. Mark, "Habits of mind: an organizing curriculum for mathematics education", *J. Math. Behavior*.

[2] R. Howe, "The AMS and mathematics education: the revision of the 'NCTM Standards'", *Notices Amer. Math. Soc.*, February 1998, 243–247.

[3] D. Schifter, editor, *What's happening in math class? Envisioning new practices through teacher narratives*, Teachers College Press, 1996.

[4] D. Schifter, editor, *What's happening in math class? Reconstructing professional identities*, Teachers College Press, 1996.

[5] H. Wu, "On the education of math majors", this volume.

[6] H. Wu, Report on the visits to four math projects, preprint, Summer 1997.

JUDITH ROITMAN
UNIVERSITY OF KANSAS
DEPARTMENT OF MATHEMATICS
LAWRENCE, KS 66045-0001
UNITED STATES
    roitman@math.ukans.edu

Contemporary Issues in Mathematics Education
MSRI Publications
Volume **36**, 1999

Afterword

WILLIAM G. MCCALLUM

The opposite of talking is not listening. The opposite of talking is waiting.

— Fran Lebowitz

Mathematicians are accustomed in their professional discourse to conditions which are alien to all other disciplines: On any given issue, there is a universally recognizable correct answer. If there is disagreement, it is because one side or the other does not correctly understand the situation. Therefore, the proper response to disagreement is to attack ruthlessly until the truth becomes clear. Once that happens, those in error will admit it gracefully and move on.

We sometimes make the mistake of expecting the same conditions to apply in arguments about mathematics education. Particularly damaging is the belief that there is no such thing as being half-right; there is nothing to be salvaged in the practices of one's opponents. Unfortunately, Fran Lebowitz's quip describes only too well much of the debate about mathematics education. One of the great pleasures of organizing this conference was to have witnessed some genuine listening. For example, the working groups on The First Two Years of University Mathematics and on Outreach to High Schools contained prominent representatives from opposite sides of the debate on mathematics education reform, yet forged remarkably unified position papers after two days of intense debate.

This is only a first step, however. In addition to listening to each other, we need to take the next step and learn to listen to voices from outside our profession.

The first group of people that we should listen to is our students. I recently had a very illuminating conversation with my daughter. Doing her homework one evening, she said:

"Dad, 14 sevenths is 2, right?" When I answered "yes", she said:

"Good, I just wanted to make sure that it wasn't division."

"But it *is* division; how else can you get the answer?"

"Oh, I was using fractions."

"How do you use fractions to show 14 sevenths is equal to 2?"

"Well... 14 is equal to 2 times 7, and the 7s cancel."

"Doesn't that mean that 14 divided by 7 is 2?" (Long pause.)

"Oh ... yeah."

My daughter is good at both division and fractions. Without this exchange, I would never have guessed that she wasn't clear on the connection between the two. Our mathematical training does not equip us to make such guesses, because it is in the nature of mathematical progress to erase the missteps in our journey towards insight. By listening carefully to our students we can eliminate the guesswork and detect the missing connections in their understanding.

We can also learn what makes sense to them and what doesn't. A business calculus class may not have much taste for mathematical abstractions, but can demonstrate great mathematical proficiency when presented with the same ideas in a concrete context that they know. The article by William Vélez and Joseph Watkins in this volume illustrates the power of presenting mathematics in contexts that mean something to the students.

The second group we should listen to is teachers and those who study teaching. (This includes, to some extent, ourselves. However, it must be admitted that university teaching has been pervaded by a dilettantish attitude which discourages serious discussion of teaching philosophy.) I include in this group anyone who takes teaching seriously as a profession; someone who can piece together with clever detective work the thinking of students, or who looks at a syllabus as more than just a list of textbook headings, or who delights in constructing homework problems that make students think about what they are doing. This group includes many school teachers and education researchers; Anneli Lax writes persuasively about what can be learned by listening to them. The group also includes mathematicians at the college level, on all sides of the educational debate, for whom, in Hyman Bass's words, "the practice of teaching has become a part of professional consciousness and collegial communication, not unlike their professional practice of mathematics itself." Jerry Uhl and William Davis in one article and Hung-Hsi Wu in another write about their college teaching experiences in a way which goes beyond personal anecdote to provide valuable professional insights.

Others worth listening to are those who use mathematics, and whose students we teach. We often invoke the opinions of engineers and scientists to defend our positions, but how often do we bother to go back to the source? I once spent a couple of hours talking to a colleague in our chemistry department. Although I obtained some useful examples for multivariable calculus, perhaps the most useful thing I discovered was the way he visualized functions of two variables. To my surprise, he did not regard the surface graph as the central geometric object, but rather the contour diagram. All his geometric reasoning about the behaviour of functions proceeded directly from the contours; it was only of marginal interest to him that the contours could be related to a surface in three dimensions. This insight fundamentally changed the way I taught multivariable calculus. The article by Dorothy Wallace gives more examples of what can be learned by talking to our colleagues in other disciplines.

Of course, we, as mathematicians, have a role as speakers as well as listeners. We must judge what are the important concepts, make sure that what is being taught is correct, and ensure a balance between technique, theory, and applications. Most of us, I think, have no trouble filling this role; it is, as Oliver Wendell Holmes said, "the province of knowledge to speak." He added that "it is the privilege of wisdom to listen." I would like to thank the many who exercised that privilege during the two days of this conference.

Reports from the Working Groups

Reports from the Working Groups

How the Working Groups Worked

The participants of the conference were divided into small working groups in the afternoons, each charged with a different topic. The task of each working group was to come up with a coherent, concise report on its deliberations, with concrete recommendations on how to improve mathematics education with respect to its topic.

Before meeting each day, participants in each working group were asked to fill out a questionnaire on their topic for the day; the answers provided the basis for that day's discussion. The group's reporter collected the questionnaires at the end of the conference.

Each group had a discussion leader and a reporter. The discussion leader's job was to keep the group working and on track. The reporter's job was to take notes, to collect the questionnaires, and to write up the final report.

The final report divides roughly into two parts, one for each day: (1) a summary of previous activities of the working group members, with an assessment of successes and failures; and (2) recommended goals and strategies to achieve those goals.

It was neither necessary nor possible in all cases that the group come to a consensus. Nor was it possible to avoid speculation and uncertainty. However, each group was advised to make its best effort at finding areas of common purpose, and recommending specific strategies.

The Renewal of Teaching in Research Departments

Members: Harvey Keynes (Discussion Leader), Al Taylor, Richard Falk, Leon Henkin, Lars-Ake Lindahl, Richard Montgomery, Dan Shapiro, Donald St. Mary, Donald Martin (Reporter), Susan Montgomery

Questions for Day 1

- What is the current state of teaching in your department? What is good or bad about it?
- What steps have you or your department taken to improve teaching?
- How does being in a research department affect your teaching?

Questions for Day 2

- What goals would you set for your department's teaching effort? (Consider curriculum, teaching practices, and infrastructure.)
- What strategies would you recommend to attain those goals? What obstacles, such as workload or reward structure, stand in the way?
- As a research department, what specific advantages or disadvantages do you have to offer your students?

Introduction. Perhaps the single greatest challenge to research departments is the renewal of teaching. Many departments are under great external pressure to do a better job teaching and to pay more attention to the needs of students heading into technical, rather than research, careers; at the same time, internal incentives remain much as they have been: slanted towards research.

The group listed the following points as basic elements in the mission of a research department:

- Research universities have a dual role of basic research and teaching.
- Every faculty member should be a good teacher.
- The measure of good teaching is that the students learn and become engaged in the learning process.
- There should be an atmosphere in the department which is conducive to students learning.
- Research departments should, in addition to the traditional work of preparing students for graduate school, provide mathematics education for the technical workforce and leadership in training of K-12 teachers.

The strategies recommended to achieve a renewal of teaching fell into three broad areas: (1) changes in faculty attitudes; (2) programmatic and curricular changes; and (3) pedagogical changes.

Faculty attitudes. Most group members felt that faculty at their institutions took teaching fairly seriously, although one mentioned a small group (most, but not all, quite senior) who 'felt that researchers need not be concerned with good teaching.' However, as one group member put it:

> The feeling of professors seems to be that most reform efforts or other ways to improve learning involve money: smaller class size, and larger teaching staff. The administration may supply some support, but not much. Some professors blame the students for inability to learn.

It was generally agreed that the reward structure needed to be changed. This was reflected in responses to the question, How does being in a research department affect your teaching?

> While being a good teacher is encouraged, it is still understood that I will basically be judged on my research output. This creates a conflict concerning time. . . .

> Not as much time available for students as one would like, and not much energy for restructuring the curriculum. Research record is the major criterion for promotions, raises, and university recognition, so there is not much incentive for really good teaching except to maintain one's self-respect, and a feeling of responsibility toward students.

> Of course the traditional structure of rewards is geared entirely toward research, and any extra time spent on teaching is discouraged. . . .

Apart from fundamental changes in the reward structure, the following measures were recommended:

- There should be more opportunities for faculty to meet students in small classes or groups.
- Faculty should be teach entire sequences with the same student group.
- There should be more regular contact between senior and junior faculty members.
- There should be improved and more extensive professional development opportunities for faculty (junior and senior).

Programmatic and curricular changes. Some departments were in the process of implementing new curricula in lower division courses, others had traditional curricula. One group member summarized the advantages and disadvantages of a traditional curriculum:

ADVANTAGES: Well-organized curricula, most people quite responsible, works OK in a traditional way.

DISADVANTAGES: Very hard to change anything. . . . Very little outreach to students; a minimal number of math majors . . . .

The group made the following programmatic recommendations:

- Programs for undergraduates not going to graduate school should be strengthened.
- More effective professional master's programs should be developed.
- Departments should make greater recruitment efforts for math majors, by offering different tracks (applied, actuarial, ...), stressing career opportunities, offering case-based courses, finding internships and undergraduate research opportunities.

Pedagogical changes. There was a consensus that experimentation with different teaching styles and with the use of technology should be encouraged, although caution was advised:

- More experimentation with teaching and pedagogy should be encouraged for some faculty.
- Effective use of technology should be encouraged at all levels of mathematics.
- There should be increased efforts to develop and use effective measures to determine if teaching approaches, alternate pedagogies, and different content are improving student learning.

The Use of Technology in the Teaching of Mathematics

Members: Peter Alfeld, Kirby Baker, Angela Cheer, Estela Gavosto (Discussion Leader), Ben Halperin (Reporter), Tom Judson, Abel Klein, Gerardo Lafferriere, Charles Lamb, John Orr, Bob Welland

Questions for Day 1

- Give a specific example of how you have used technology in your teaching (Which courses, how much use, who used it, what was it used for: motivation, illustration, heuristic arguments, numerical computations, symbolic computations, graphical work, more advanced problems).
- How successful was it? How did it help or hinder student learning, compared to teaching the same topic without technology?

Questions for Day 2

- In general, what are the problems to think about when considering the use of technology? (Effect on later courses where technology is used, training for the workplace, ...) Distinguish between problems that are inherent in the technology and those coming from the implementation.
- What do you think the future role of technology in teaching mathematics should be?

Introduction. The discussions in our group were very lively during the two days that the group met. The dominating concern was that technology is already a part of everyday life. In the near future, whether we like it or not, it will be an essential part in the teaching of mathematics. The choice for research mathematicians is what role we will play in this change. In our group, we covered a broad range of issues. Not surprisingly, the computational capabilities of the technology were not discussed much. Main issues were the potential of the technology as a communication tool and the demand for it in the workplace.

Recommendations. We want to urge all mathematicians to take an active role in the use of technology in the teaching of mathematics. We are the only ones that have the mathematical knowledge and the teaching experience to make it worthwhile.

We should take a realistic (optimistic) approach about students' abilities and computer availability. There are two main ways that we think that the technology can be productive in the teaching of mathematics:

- as a computational tool and
- as a communication tool.

The big questions about the computer as a computational tool (numerical and symbolical computations and graphics) are related to the students' learning. In particular, the following questions should be addressed:

145

- How can technology be used to further students thinking beyond the traditional contents of a course?
- How can technology be used to teach "remedial" mathematics like deficiencies in algebra?
- How can technology be used to help student develop independent visualization skills?

We are witnessing an explosion in the increase of the use of technology as a communication tool. Electronic mail is a very powerful way of facilitating communication among students and instructors. The use of the web is in exploratory stages. Some of its possible uses are the following.

- Online Courses
- Traditional courses:

  - web course notes

  - discussion groups

  - on-line office hours

  - pre-testing and testing

  - on-line demonstrations

Physical considerations should also be a very important consideration from the students' perspective. Actually, there are very different environments where technology can be used:

- the computer laboratory;
- the traditional classroom with equipment for computer demonstrations;
- the traditional classroom with portable technology like a graphing calculator or calculator based laboratory; and
- web–based activities.

All these different environments seem to converge. A primary obstacle for the implementation of any new technology is the training of the instructors. Serious consideration and effort should be given to this matter.

We want to conclude by saying again that it is imperative that research mathematicians participate in defining the technology that will be used to teach mathematics in the near future.

Different Teaching Methods

Members: Greg Baker, David Epstein, Ted Gamelin, Sid Graham (Reporter), Ole Hald (Discussion Leader), Delphine Hwang, Suzanne Lewis, Randy Mc-Carthy, Brad Shelton, John Sims, Robert Underwood

Questions for Day 1

- What is your favorite teaching method? (Standard lecture method, collaborative learning, labs, etc.)
- What experiments have you tried in order to improve your teaching methodology? (Getting students actively involved in the lecture, group work, lab projects, using technology, etc.)
- What did you want to achieve with your experiment? Did you achieve it? Were there any unexpected results?

Questions for Day 2

- What goals would you set for alternative teaching methods? (For example, getting students to read, write, think for themselves, take responsibility for their work.)
- What strategies do you recommend to achieve these goals? What obstacles stand in the way?

The participants in this working group exchanged ideas on teaching techniques that they either use regularly or with which they have experimented. Over time, enthusiastic mathematics instructors have attempted various new and old teaching techniques. The wheel is constantly being reinvented. Nevertheless, the characteristics of the body of students taking calculus are constantly changing, teaching resources are changing, and teaching techniques require constant experimentation and adaptation to new circumstances. It is a worthwhile exercise to compile an assortment of teaching techniques, and to describe their assets and pitfalls. We begin here to create such a collection.

We describe here several of the techniques that various committee members have used in recent years. The techniques we list are all designed to work in large lecture sections; many of the techniques are adaptable to the small class format as well.

Summary of previous activities

Advisory Committee of Students, or Ombudsmen (Randy McCarthy). This technique has been used by Randy McCarthy to deal with the problem of communication between students and instructor in large lecture classes. He forms an advisory committee of students, which meets regularly with the instructor to communicate the students' concerns. The advisory committee students — or *ombudsmen* — have regular communication with individuals or groups of students, and make it their business to be in synch with the needs and feelings of the class. Students open up to their peers on topics which they would be most uncomfortable discussing directly with the instructor. The ombudsmen develop a good working relationship with the instructor, and the resulting good feeling permeates the rest of the class.

Use of Worksheets and Breaks in Large Lecture Classes (Ole Hald). Ole Hald has used a version of this technique at the University of California at Berkeley. For a calculus class of about 400 students, of which 300 might be present at any given lecture, about 100 worksheets are distributed very quickly at the beginning of the class by a large number of assistants. Hald might lecture for ten minutes and then direct the students to the first problem on the worksheet. For instance, he might show them how to integrate $\cos(x)$, and the worksheet problem might be to integrate $x + \cos(3x)$. The students work in groups of three or four. Though the sheets are not handed in, the students are asked to put their names on the top of the sheet, and this way they learn about each other. If the classroom is large enough, students are asked to leave every third row empty, and then Hald can circulate around the room and reach all students. After ten minutes on the worksheet, there might be another short lecture segment, then back to the worksheet.

Use of TA-administered 15-minute oral exams to cover theory (Ole Hald). Ole Hald is experimenting with this technique at U.C. Berkeley. The main idea is to dedicate a week (or four days) of the schedule to oral exams on theory. The students are given ten exam topics to prepare. They sign up for exams at 15-minute intervals. The exam is administered by a TA, and another TA is present as a scribe. At the beginning of the exam the student is offered two topics and selects one. Then the student is given eight minutes (no more!) to present the topic (such as an easy theorem about convergence of series and its proof from the text). In the next few minutes the student responds to questions from the proctors; these questions are designed to test understanding. The student leaves, and the two TAs assign a grade. The grading is on a scale of ten points. For a perfect presentation, the student is allocated eight points, and the student's score fluctuates up or down from this base level by at most two points according to the responses to the questions. (Thus, to guarantee a score of six points, the student must simply to commit to memory the presentations of nine — out of ten — theorems and proofs.) At the beginning of the exam the student is permitted to reject the two topics offered and ask for two others, at a cost of two points deducted from the final score. How much total TA time this requires depends on the size of the TA sections (not the class size). It does require extra TA time and intensive effort during the oral exam week. No classes are given during that period. The instructor gives some training to the TA's before the oral exams, and the instructor visits exams randomly. The student is allowed to bring a friend as witness to the exam, and some students have brought teddy bears to their oral exams.

The two students participating in our workshop session, Delphine Hwang and Suzanne Lewis, have taken these oral exams. They feel that the method is effective. They practiced their presentations with each other before the exams and they learned a lot in the process.

Hald has also used variants of this technique in smaller, more advanced undergraduate courses.

Use of Summary Statements and Paragraphs (John Sims). The idea is to have the students write summary paragraphs of individual sections and chapters of the text. This forces students to think through the material on their own and to organize their own synopses. It helps students to prepare for exams. One way to implement the idea is to give the hour exams in two parts. One part is given in class; the other part is a take-home exam that includes a request for the written summary. The summary might be three sentences long, and the concatenated summaries give a good overview of the course.

Use of Crib Sheets. Crib sheets are very brief sets of notes taken into an exam. This is a widely-used technique; it has many variations. It also encourages the students to come up with a short overview of the material in their own words. Some instructors prefer to distribute the cards for the crib notes, and even to use colored cards so that they are easily recognizable in exams. Others allow the students to use their own materials. There are widely differing views on whether students should be allowed to use calculators in exams.

Selective Grading of Homework. In an era of diminishing resources, it is often impossible to read all the homework of all the students. Yet it is very important that students do the homework, and on a regular basis. And it is important that they get regular feedback on their work.

One compromise is to require that the students turn in substantial homework, and have the reader grade only a portion of the homework problems. If there are not even resources for this option, then it is better to require the students to hand in homework and simply check it as done than to give the students the responsibility for doing the homework and checking their answers against some posted answer guide.

One method of selecting homework for grading, used by Ray Redheffer at UCLA, is to select which of several assignments will be handed in for grading by a random choice, as follows. Ray gives two homework assignments in every class meeting. He has the students bring the two completed assignments to class in the following lecture. At the beginning of the class Redheffer asks three students to spell their last names, which he writes on the blackboard. He counts the total number of letters in the names, and the parity determines which of the two assignments is to be turned in. A byproduct of this method is that Redheffer learns the names of a number of students in his large lecture.

Another method of solving the reading problem has been used in large calculus courses at UCLA by Bob Brown. While no homework is handed in, there is a weekly quiz in recitation section consisting of exactly one homework problem from a designated list. The problem is chosen at random, and it might (or might not) be different for each recitation section.

Use of Mastery Exams (David Epstein). These are exams given during the term that a student must pass at a level of 100% in order to pass the course. The course might have two midterms, a final, eight quizzes, and six mastery exams. If the students passes all the mastery exams then he/she is guaranteed a grade of C in the course. Grades on midterms and other parts of the course then serve to raise the student's aggregate grade from the entry level of C.

The mastery exams are designed to be straightforward and simple, and the student may repeat each mastery exam as often as necessary. Of course the professor will require considerable help (from TAs or other assistants) to administer such a system.

Recommendations. The working group recommends that an appropriate education committee of one of the professional societies maintain a Web site that would serve as a repository for descriptions of teaching techniques, including details of implementation, commentary on hidden difficulties, what works, and what does not work. The Web site should also maintain a bibliographic listing of books or articles on teaching techniques.

The First Two Years of University Mathematics

Members: Joseph Ball, Christopher Grant, Peter Lax, Robert Megginson (Reporter), Kenneth Millett, Wayne Raskind, Thomas Tucker, Joseph Watkins, Hung-Hsi Wu (Discussion Leader)

Questions for Day 1

- Describe the composition of the freshman and sophomore mathematics class of your institution. (Math majors, science and engineering students, business students, premeds, general education students.)
- Have you undertaken any projects involving freshman and sophomore courses? Consider both curriculum and teaching practice.
- What would you change if you undertook the same project today?

Questions for Day 2

- Consider the factors that affect your department's ability to do a good job teaching these courses (workload, homework policy, grading practices, availability of technology, student preparation, faculty commitment, curriculum). Which ones can your department affect, and how can it do so in a systematic way?
- What strategies do you recommend to achieve a mathematics program that your department teaches well, and that satisfies the standards of both your department and the home department of your students?

This working group endeavored to discover the distribution of students by major in the first two years of mathematics at the institutions represented by the working group members. It also discussed what projects are being undertaken in freshman/sophomore level courses at these institutions and what has been learned from them, and what strategies would help the departments achieve a well-taught mathematics program for these students.

The distribution of students by major in these courses varies greatly from institution to institution. None of the working group members reported more than about 20% mathematics majors in these courses, and one institution reported only 2%. Though not all members were able to report figures, an estimate for the median percentage of mathematics majors in these courses at these institutions would be about 10%. The distribution of non-mathematics-majors in these courses varies greatly depending on the mission of the institution. For example, in one institution about 50% of the Calculus I students are premedical, while another has no such students because it has no premedical program. It is clear that, looking over all these institutions, there is no "typical" distribution of students — according to major — in the first two years of mathematics.

Almost all of the institutions represented by this working group are doing some sort of experimentation with or modification of their mathematics courses for the first two years. Some of the projects are significant across-the-board calculus reform efforts, while others involve smaller modifications to individual sections of a course by the instructor teaching the section; there are yet other efforts that are a mixture of the two.

Summary of previous activities. Virginia Tech has three paths through calculus, with each path appropriate for a different group of disciplines. Some sections of differential equations courses have been taught for individual disciplines. *Mathematica* is used regularly in engineering calculus. There have been experiments with self-paced, interactive, computer-based instruction in precalculus. There is also an Emerging Scholars Program involving mandatory extra problem sessions — to give students with marginal backgrounds a better chance of success.

In the past several years, BYU has had three strands of calculus: one taught from the Harvard Consortium text, one from the `Mathematica`-based text by Stroyan, and one from Ellis & Gulick. The teacher of each section is free to choose one of the texts according to personal preference, and no differentiation between the courses taught from these various texts is made in the timetable. Starting in the second term of the 1996–1997 academic year, all students in calculus will take a common competency exam. "Teaching trios" are being introduced, each consisting of a senior faculty member, a newer faculty member, and a graduate student; these three team members visit each others' classes. A course has been implemented that is designed specifically to teach students about proofs.

NYU has integrated applications extensively into its calculus courses. Currently, full use of graphing calculators is being made in these courses. There has been some experimentation with the Harvard calculus materials. While the instructors liked them, this feeling was definitely not shared by the students.

The University of Michigan has large-scale calculus and precalculus reform efforts underway, and uses the Harvard Consortium's precalculus and calculus texts in all regular sections of their Precalculus, Calculus I, and Calculus II courses. Group methods and graphing calculator technology are used extensively, and a week-long training program at the end of August prepares the instructors to teach these courses.

UCSB has two calculus tracks, one for the mathematicians, scientists, and engineers that is taught from the Harvard text, and one for the life scientists and economics majors taught from *Calculus in Context*. There has been quite a bit of experimentation with pedagogy in these courses. Currently, the following courses are undergoing renovation and development: Mathematics for Elementary Teachers; a course for majors to aid the transition to upper division courses; and problem solving and history courses for future teachers. There is also a substantial workshop program, associated with the traditional "barrier" courses, directed toward increasing the participation and success of women and underrepresented students. With the passage of California Proposition 209, there is increased pressure threatening such programs at UCSB, especially in mathematics. UCSB is also currently in the midst of a counter-reaction to efforts to increase the effectiveness of the program for the first two years; resulting negotiations have extended the scope of the discussion to bring in faculty members who have seldom, if ever, taught many of these courses.

USC has experimented with reform in the past, and the working group member from USC, Wayne Raskind, is a member of the Harvard Calculus Consortium. The current main effort at USC is along traditional lines, using the Stewart text. There are plans for an enhanced calculus course for students who come in with AP credit, with some linear algebra included in the first semester of this course. Some efforts are being made in the direction of group learning.

While Colgate has no large-scale projects to change courses in the first two years of mathematics, various instructors are experimenting with changes in individual sections. A few are using weekly computer assignments, and a few have used the Harvard Consortium text. (Two years ago, the department agreed on a core syllabus for Calculus I and II, but instructors are free to choose their own texts. There are common final exams.) All sections use graphing calculators, but there is no formally prescribed way in which they are to be used in the courses, so much of their use is similar to that of traditional calculators.

Arizona has had a major presence in the Harvard Calculus Consortium; several of it faculty are members of that consortium. The Harvard texts are used for the first three semesters of calculus. There is a training program, lasting several days, for those new to the teaching of these courses. On occasion, some sections have used other texts and teaching styles. The Mathematics Department has recently designated calculus and precalculus "czars" whose job is to provide institutional memory and consistency for the calculus and precalculus courses. The role of the precalculus czar is particularly important at the moment, since the precalculus program is changing in response to increased entry requirements for the Arizona university system.

The finite mathematics and calculus courses for business students at Arizona are moving toward the use of spreadsheets as part of the pedagogy of those courses. Two consortium members have written a text for differential equations, and technological ideas are being added to the linear algebra course. One general difficulty currently affecting teaching assignments is a mandate from the Board of Regents that 2/3 of all freshman and sophomore teaching be done by faculty with Ph.D.s.

About two years ago, the chair of this working group, along with a colleague, reorganized the first two years of calculus at Berkeley. The principal changes were to eliminate linear algebra from multivariable calculus, to use the same book (Stewart) for three semesters, and to make the two sophomore courses (multivariable calculus and ODE–linear algebra) independent of each other. There is another effort underway, again spearheaded by the chair of this group with another colleague, to have the "soft" calculus sequence at Berkeley use the Harvard Consortium's *Applied Calculus* text.

Recommendations. Much has been learned from the above projects. There were comments from several members that assessment is a critical issue, and one that would perhaps be better thought out if some of the projects were to be undertaken anew. It was observed that all too often departments make significant changes in programs or texts without adequate procedures in place to assess the results. Agreement should be reached ahead of time on such issues as what needs to be improved, what the goals of any proposed reform are, what the strategies are to reach those goals, and other related issues. (In several of the universities represented by this group, assessment of reform efforts has shown no discernible difference between test scores of students coming from traditional and reform sections, but it was observed that this can have different interpretations depending on the design of the tests. The scope of assessment can be fairly broad, and should probably take into account aspects of student self-esteem and empowerment.)

It was also observed that the workload needed to support curricular reform projects can be great, and that in retrospect this could be taken into account to a greater extent. Another observation was that more communication with client departments about the nature and extent of the reform efforts could in certain cases have been useful.

The working group's discussion of possible changes to mathematics programs in the first two years focused on three issues: **(i)** the role and form of linear algebra in the first two years, **(ii)** the role of abstract mathematical reasoning in introductory courses (by whatever name one wishes to attach to it, whether *proof,* or *rigor,* or *theory*), and **(iii)** the need to keep lines of communication open to client departments. No agreement was reached on the role and form of linear algebra, with some preferring that it be strongly motivated by applications and computation, and others wishing to see more of the structure and theory of linear algebra introduced.

There was general agreement that, in the first two years, students should be shown the need for rigorous mathematical argument in appropriate situations, rather than being exposed only to heuristic arguments. There was also general agreement that the instructor should always seek the best way to convey *understanding* of an idea, result, or technique; this goal may at times require proof and at other times call for less rigorous arguments based, for example, on graphical intuition.

Some members of the working group felt that mathematically correct proofs of several main results should be given in all introductory mathematics courses, to the extent of perhaps four or five such proofs per semester; also that the students should be shown how to construct rigorous mathematical arguments themselves, lest they suddenly discover that they are not properly equipped to continue in mathematics when they reach the more proof-intensive advanced courses. Others felt that this is not so important in the first courses, but that instead appropriate opportunities could be sought to develop in the students' minds an understanding of the need for proof and generalization in arguments. One example that was given involves first showing students that $x/e^x \to 0$ as $x \to \infty$, then doing the same for x^2/e^x, x^3/e^x, and x^4/e^x. This investigation does provide heuristic evidence that $x^n/e^x \to 0$ as $x \to \infty$ whenever n is a positive integer. However, the students might feel uncomfortable with the generalization since, for example, x^{1000} grows so much faster than x^4. In addition, after seeing the four special cases, the students will probably conclude that it would be good just to do this once rigorously for x^n/e^x and be done with it.

The working group also discussed the need to design courses to meet the needs of client departments, and possible mechanisms for keeping lines of communication open with those departments. It was recognized that one common difficulty involving lines of communications is that modifications can be made in consultation with certain persons in client departments who may have gone on sabbatical or may not be in the appropriate roles to influence the process at the time the changes are actually implemented. It was suggested that stability can be achieved by having some persons in long-term roles to make decisions of this sort and to interface with their counterparts in the other departments. Some members also felt that efforts to assure long-term stability in courses — in particular to standardize course content and classroom pedagogy over a long time period and over multiple sections of a course — bump up against serious issues of academic freedom; and this tends to be a force against such stabilization.

The Mathematics Major

Members: Jorgen Andersen, John Brothers, Ralph Cohen, Stephen Fisher, Andrew Gleason, James Lin, Lea Murphy, Richard Montgomery, Y. S. Poon, Ken Ross (Reporter), Anthony Tromba (Discussion Leader)

Questions for Day 1

- Are you satisfied with the experience for mathematics majors at your institution? In what ways have you tried to improve it? (Tracks for different professional interests, undergraduate research projects, math clubs, mentoring, capstone courses, summer internships.)
- What worked and what didn't?
- Do you have many students from other departments taking your upper division courses? How do these students compare with your majors?

Questions for Day 2

- What should we be trying to provide most of our math majors: training for graduate school, training for professional careers, a general education in mathematics?
- What strategies do you recommend for improving the education of math majors? At what stage should majors be introduced to proofs?

Summary of Previous Activities. It has been the case historically that mathematics majors have been produced by schools of every type, from liberal arts college to technical institute. Any discussion of the mathematics major should take into account the various types of mathematics majors that there are, and where they are trained.

Here are some general trends concerning the major; these are distilled from questionnaires that we distributed among the panel members:

- Tracking in the mathematics major, also known as a choice of "options", is a common device for helping the student to have a focus for his/her studies. A typical example of a track is an "applied mathematics track". It consists of a specific curriculum of courses and activities that will help an undergraduate student to train as a mathematics major with applied skills. Such a curriculum will contain some specific mathematics courses, such as numerical matrix theory and partial differential equations; it will also include courses outside the department, such as computer science, physics, and engineering courses. Other standard tracks include "computational math track" or a "biomathematics track" or a "statistics track" or a "mathematical physics track".
- Student math clubs, such as $\pi\mu\varepsilon$, are often unsuccessful; this is so because they are dependent on the involvement of a dedicated faculty member and on a few key students to act as catalysts. The math club can be more successful if some of the meetings are career-oriented and if alumni are involved. In

particular, alumni can provide the students with valuable career information, and also give living credibility to the notion that mathematics is something that real people use in the real world.

- The trend in other departments such as engineering, biology, and chemistry has been to increase the number of course requirements needed to complete their majors. The net result is that students from other departments have less time than perhaps they once did to take additional mathematics courses.

- Surprises included: Some of the best students in upper division courses at colleges represented on our panel are in fact high school students; at two of the schools represented on the panel, the math majors comprised 0.5 percent of the student body. It has been a nationwide trend that the number of science majors in general, and the number of mathematics majors in particular, has fallen steadily over the last decade or more. Reasons for this decline include

(i) Student interests have moved in new directions, so it is less natural for the brightest students to go into math than it may have been twenty years ago.

(ii) Many students today graduate high school not having the study skills and the intellectual values that students were expected to have twenty or more years ago. As a result, even if they come to college thinking that they want to major in mathematics, they find that they cannot cope with the workload and the intellectual depth and they consequently switch majors.

(iii) Students find mathematics to be dry, uninteresting, and not engaging. They do not have a clear picture of what they will do with a mathematics degree once they graduate (this despite the fact that more and more fields are becoming mathematized).

(iv) Concomitant with (iii), students are easily enticed by the ready success and easy entree offered by majors like business. Traditional subject areas on the whole, and mathematics in particular, have suffered a loss of student interest and therefore a loss of majors.

The use of technology was discussed. Maple, Mathematica, MatLab and other computer algebra/graphing systems have been used successfully in calculus and upper-level courses without detracting from the main goals of the courses. Students often experience some initial discomfort in learning the syntax of one of these languages and in becoming inured to formulating mathematical questions in the computer environment. Once they internalize the language, then this problem begins to fade away. The curriculum is especially effective if students see the same computer language in several courses.

A basic question we should ask ourselves is, "What do we want technology to do?" Do we use it to illustrate concepts, sometimes using vivid graphics? Do we want it to take over tedious calculations so that students can concentrate on higher order ideas? Do we want the computer to perform massive calculations that would be infeasible to do by hand? Do we want to use the computer as a modeling tool? Or is it to be a primary learning tool? Is learning to use software part of students' training to make them more marketable? We theoretical mathematicians often find ourselves unfamiliar with this territory; but in today's climate it is essential that we become conversant with the relevant issues.

Recommendations. Mathematics majors should be shown both the intrinsic beauty and the applicability of mathematics. They should also understand that mathematics is a living subject with a fascinating history that has made substantial contributions to our culture. Such considerations deserve to be a part of all mathematics courses at every level.

With the introduction of technology and also more applications in our courses, some other topics will have to be eliminated. We cannot cover all of the old topics and cram in the new and expect to teach coherent courses. Many of our current sophomore-level courses used to be taught in the junior year, so our expectations have been raised at the same time that the students' skills seem to be dropping.

It is also the case that many of the new teaching techniques—group learning, self-discovery, and computer labs are just three instances—use time in ways that are unfamiliar to many of us. The more traditional lecture teaching method is a highly structured didactic device which gives the instructor complete control of the use of time. An instructor inured to that classical methodology will, in effect, have to be retrained so that he/she can use the new techniques effectively and well.

Mathematics majors must know the difference between a precise statement and a vague assertion; some in the working group felt that every undergraduate mathematics major should take a course with serious focus on proofs and should be able to write a proof of some sort themselves. We noted that there is a difference between (i) understanding the ideas and (ii) studying the proofs in detail, or even in understanding the details. Getting lost in the epsilons and deltas and missing "the big picture" is all too common and very undesirable. Equally unpalatable is the prospect of a student getting the big picture but not understanding the inner workings of the subject. A good mathematics course is a balancing act among abstraction, careful rigor, precision, and problem-solving.

It is also critical for graduating mathematics majors to have been presented with the history of the development of ideas in mathematics. There are important ideas and important theorems that do not develop in a vacuum and which must be seen as part of the natural development of the subject.

We recommend that more attention be given, both in courses and in textbooks, to the broader context into which any particular idea fits. To achieve this goal, instructors may have to decide to prioritize what must be proved and consequently which proofs must be sketched or in fact omitted entirely. Furthermore, it should be made clear that commonplace mathematical ideas were (historically) often achieved only after a long and difficult period of development. For example, progress in understanding basic concepts like the real numbers, limits, and continuity was very slow. Students should come to understand the reason for the development of abstract and axiomatic approaches.

This highly interactive melange of ideas should be kept in mind when departments are revising syllabi and creating new courses. A new course is not simply a list of topics. It is a symbiotic collection of ideas, the body of which fits into an historical and a scientific context. The course should also fit rather naturally into the mathematics curriculum; and that role of the course should be apparent to instructor and student alike.

The Education of Non-Mathematics Majors

Members: Adeniran Adeboye, Stephen Greenfield, Jean Larson, Ashley Reiter (Reporter), Dorothy Wallace (Discussion Leader)

Questions for Day 1

- What projects have you been involved with regarding the education of non-majors? (Consider individual courses, such as calculus for scientists, engineers, and general education courses; and consider larger projects, such as interdisciplinary projects with other departments.)
- What worked and what didn't?

Questions for Day 2

- What mathematics should the average student know? (Algebra? Basic Probability? Geometry?)
- What is the appropriate balance of drill and conceptual understanding for the average student? (Algebraic skills, ability to analyze data in a graphical or numerical form, applications of mathematics, etc.)
- What strategies do you recommend to teach students the knowledge and skills they need?

Much of the discussion of undergraduate mathematics focuses quite naturally on the needs of mathematics majors. However, non-math majors make up the overwhelming majority of the undergraduate students taught by departments of mathematics at U.S. colleges and universities. We should think of these students when we analyze our curriculum and our course offerings. The purpose of this report is to initiate a discussion centered on those students whose need for mathematics is directly related to their study in some other subject or in order to satisfy a degree requirement.

Summary of discussion. We think that students interested in physics, chemistry, engineering, computer science, and biochemistry need skill in mathematical modeling and the ability to use mathematical techniques outside of mathematics classes. A more recent development is that psychology, economics, earth sciences, nursing, business, and other majors who are not from a core science are also required to learn some mathematics. At many schools there is a quantification requirement, thus in effect mandating that every student at the school take some mathematics. The mathematics department should be prepared to rise to this challenge, and to fill the need. Apart from abstract philosophical reasons that can be mounted to justify such an action, it is also crucial for the survival and well being of the mathematics department that it be prepared to meet any and all curricular math needs on campus.

The students who come to us for service courses should be able to recognize when a problem in their field calls for a mathematical approach and then to follow the problem from the original considerations to a mathematical description (including some consideration of the simplifications and errors that modeling may introduce). They should then understand the predictions of the model, and use those predictions to check the validity of the model. They should also understand the precision and power of mathematics and its logical structure and cohesiveness. All of our service courses should be designed with these needs in mind. Our sensitivity to student and to curricular needs may get us more "converts", and it also should make the students more reliable and appreciative consumers of mathematics.

The needs of students of biology, nursing, psychology, economics, earth sciences and business may be a bit different from those in the hard sciences. The qualitative needs described in the last paragraph will still apply; but a business student will want to learn some mathematical modeling, qualitative calculus ideas (rates of change and numerical approaches, phase portraits, etc.), linear algebra, statistics, and some areas of discrete mathematics. This mix of basic analytical techniques will serve these clients better than some more traditional subjects.

Future teachers also have particular needs. Prospective teachers should learn mathematics that goes well beyond what they will teach, but that is salient to what they will teach. We also need to help teachers develop mathematical ideas and provide them with mathematics that they can use in their classrooms, while modeling pedagogy appropriate for their use.

We noted that liberal arts students often must fulfill a quantitative skills requirement. This population is large and growing, and may not be well-served by the type of precalculus course now being offered. A history or literature student may want to know some mathematical history and some mathematical culture; that student's need for analytical problem solving skills, or for integral calculus, or for vector analysis will be much less.

Pre-professionals need certain skills for exam preparation (e.g., the MCAT, LSAT, and GMAT). The mathematics department would be well advised to become acquainted with those exams. Consultation with relevant departments is also recommended.

Recommendations. Clearly the mathematical needs of non-math-majors will vary significantly with the university and academic major. Each department must therefore do its own analysis. Communication with faculty members in other departments and other programs can be extremely valuable in finding out what students in other majors actually need. We strongly urge that math departments engage in a dialogue with representatives of other faculty groups to try to discover what is needed and how to fulfill these needs most efficiently. Important questions to consider are:

- What is the mission of your institution?
- Who are the students in your classes (with respect to major, career track, graduation requirements)?
- For each major, what math is specifically used and where?

While the particular ways in which non-math majors can be best served will vary from institution to institution, the following general recommendations may be appropriate at most schools:

- It would be ideal if individual senior faculty were responsible for specific non-major courses for a period of several years. Each course stewardship should have a duration of (about) five years. It would also be ideal if the "ownership" of a course changed on a regular, scheduled basis. The purpose of this system is to facilitate communication with other departments and interest groups, as well as to develop among faculty a sense of responsibility for the education of non-math majors.

- Too often in the past have we funneled non-majors into some form of calculus course. In many instances this choice fit poorly the needs of our clients. As noted above, many non-majors—business majors and psychology majors come first to mind—could make good use of some statistics, some finite math, some problem solving skills, and *certain parts of calculus*. We should endeavor to design service courses that actually fit the needs and values of our clients. Service courses should be regularly re-examined to insure that they are kept timely and continue to meet the needs of the departments and the students they are intended to serve. This review should be conducted in consultation with client departments.

- All courses, regardless of content, should be presented with a view towards communicating something of the mathematical endeavor. This can be done in many ways. For example, we could give a historical perspective, showing what sorts of questions mathematicians have asked, and what efforts have been made to answer these questions.

- When a service course is designed and implemented, the language and the intellectual values of the client disciplines should be kept in mind. An effort should be made to guarantee that the course is both valuable and credible to the students who will be taking the course.

- Pedagogy in recent years has responded to changes in generally available technology and to changes in educational style. Graduate students, as prospective college teachers, should be helped to respond effectively to these opportunities. They should be made aware of the many new teaching techniques that are being developed, and also of the many technological tools that now exist. Graduate programs in mathematics have a responsibility to their students to enable them to teach effectively.

Departments may have difficult debates when dealing with some of these suggestions. Change is rarely easy and painless. We therefore further suggest:

- Maintain civility and respect for colleagues across disciplines and within the math department.
- Build on existing friendships and camaraderie when possible.
- Allow each teacher more flexibility to teach each section in a way which is consistent with his or her own philosophy, at least until consensus is reached.
- Be aware that it is quite difficult for an individual faculty member, regardless of age and level of experience, to analyze his or her own teaching and to master new teaching techniques. A departmental system of mentoring and faculty development should be formed.
- Resolution of outside requirements and departmental needs is worthy of detailed discussion.
- Be aware of political pressure points at your school (again, ask client departments) which can be used to help in developing the service curriculum.
- Some external support, such as a supportive dean, funding to purchase hardware, or released time for faculty to plan or restructure courses may be very helpful in this transition.

The teaching of service courses has long been the "silent partner" in the mathematics enterprise. The stunning growth of mathematics departments in the 1960s was due in large part to the need for more faculty because of substantially increased enrollments in service courses. We need to take a more active role in developing service courses that accurately address the needs and curricular requirements of the departments that we profess to serve. If we do not do so, and in a credible way, then our clients will begin to develop their own mathematics courses; such a development would be to our detriment. In the long run, well-thought-out service courses will be important to the health of mathematics.

Outreach to Other Departments

Members: Chris Anderson, Barbara Bath, Marjorie Enneking, Terry Herdman (Discussion Leader), Paul M. Weichsel (Reporter)

Questions for Day 1

- What students and what departments do you serve in your mathematics courses?
- To what extent have you worked with other departments in the design or delivery of mathematics courses? (Engineering, Sciences, Business, Social Sciences)
- Which projects were successful and which weren't?

Questions for Day 2

- What strategies do you recommend for finding out the needs of other departments?
- For each group of students that you serve, what mathematics do you think they need?
- What strategies do you recommend for setting up successful, sustained joint activities with other departments, and which activities do you think are likely to be most fruitful? (Joint committees, regular consultation, finding out how they use mathematics, team teaching, interdisciplinary courses.)

Outreach has been a fact of mathematics department life for the past forty years. But changes in the university structure, in societal values, and in the needs of the student body demand that the mathematics department take a more active role in developing and nurturing its outreach activities. This work group report explores some of these new needs.

Basic principles.

I. *Outreach must be based on a notion of* ***partnership*** *rather than a client-provider relationship.* There should be regular meetings between mathematics department representatives and client department (physics, engineering, psychology, etc.) representatives in order to determine the shape and content of any given service course. An effort should be made to have some continuity in the membership of these interface committees. Since different departments often have entirely different vocabularies and value systems, real effort must be made to open lines of communication. Committee members should examine texts and discuss syllabi. Department chairmen should have a role on these committees.

II. *Lines of communication with other units need to be kept active on a continuing basis whether or not both units are engaged in a specific joint project.* There should probably be twice-yearly meetings just so that the departments can touch base with each other and assess the status of courses that are in place. This is also a time when new needs and ideas can be broached.

III. *Mathematicians need to be prepared to invest the time and effort necessary to learn enough about other disciplines and help others to learn about ours in order to make interaction fruitful.* While the responsibility for communication lies on both sides of the fence, we must bear in mind that *we are the serving department*. The mathematics department's health and well-being depends on how well we are *perceived to be fulfilling our service role*. Therefore we must be prepared to go the extra distance to communicate with our clients.

Recommendations and examples.

(i) *Engage in interdisciplinary research which may lead to interdisciplinary educational activities.* Such collaboration will not only be mutually beneficial on a scholarly level, but will help faculty to develop the common vocabulary, and also the trust and rapport, that is essential to any kind of cooperative venture.

(ii) *Work with other departments to develop joint majors/minors and double majors.* In today's climate, the pure mathematics major is playing an ever smaller role in the educational milieu. Collaboration with other departments will help the mathematics department to develop viable programs in statistics, biomathematics, mathematical physics, and other applied disciplines.

(iii) *Examples of course development:*

- Team-taught courses. Ideally, the "team" should consist both of math department members and of faculty from other departments. The subject matter should be of interest to students and faculty from both fields, and also to faculty in departments across campus. Team-taught courses are an effective device for increasing campus appreciation of the mathematics department and the contribution that it makes to the educational effort.

- Courses jointly developed by several departments and taught by one. Examples might include a course on applied partial differential equations, one on general relativity, and one on mathematical genetics.

- Custom courses designed for special audiences. This could include courses in acoustics or in wavelet algorithms or in applications of C^* algebras to physics.

- General education courses. These would be courses for non-majors, often students majoring in the social sciences or humanities who need a "quantification" course for a breadth requirement. It will also include K–12 teachers.

- Work with other departments to ensure the mathematical integrity of their courses. This activity has the usual pitfall that the participating mathematician(s) might be seen as "talking down" to their clients. Care should be taken by those who are trying to help other departments to make their courses mathematically honest.

- Work with other departments to make other courses appropriate for math majors. The same caution as in the last item applies. This can be a fertile field of interaction with other departments, as obviously those departments want to attract more students (particularly math students) to their courses.

(iv) *Establish a unit consisting of faculty on temporary assignment from their home departments to jointly develop and teach interdisciplinary courses.* This is a sound idea that will obviously require administrative support — in the form of release time, or flexible scheduling, or support staff. It may also require pecuniary resources — for equipment or staff or guest speakers.

(v) *Campus-wide committees to engage in course revision over an extended period when relevant.* Mathematics plays a central role in any College of Arts and Sciences. As we approach the millennium, mathematics impinges on all sciences and on many other subjects as well. Campus-wide committees help to institutionalize the pivotal role of mathematics and also help to open up lines of communication both with client departments and with the administration.

(vi) *Work cooperatively with other departments on software and hardware decisions.* Most science departments outside of mathematics do not use TeX. Only some use `Mathematica`; many instead use `MatLab`. Physics departments still use `Macsyma` in certain applications. It would be well for us to become aware of how other departments use technology; such knowledge is certainly essential in designing new courses and new curricula.

Traditionally, little formal effort has been expended by mathematics departments in trying to develop relationships with other departments. In those happy circumstances where a productive relationship did develop, it was usually through the serendipity of a few individuals. It would be well to institutionalize the lines of communications between math and its allied physics, engineering, premed programs, and so forth. Clearly the long-term health and prosperity of the mathematics enterprise is increasingly dependent on such communication. Institutional support for these outreach activities is essential — including adequate reward and appropriate recognition.

Outreach to High Schools

Members: Gunnar Carlsson, Phil Curtis, Dan Fendel, Neal Koblitz, Anneli Lax (Discussion Leader), Judith Roitman, Tom Sallee (Reporter), Martin Scharlemann, Alina Stancu, Abigail Thompson, David Wright, William Yslas Vélez

Questions for Day 1

- What work with high schools have you been involved in? (Visiting classes, in-service, co-op programs, programs for high school students, summer courses, faculty volunteering in high schools.)
- Which were fruitful and which weren't?
- Do you have any programs for teacher training at your institution?

Questions for Day 2

- What do you think is the appropriate role for university faculty in the development of high school curricula and teaching methods?
- What strategies do you recommend to improve the communication between local high schools and your institution?
- How do you think your institution can help high school students learn mathematics?

It is too easy for university mathematics faculty to bemoan the preparation and the attitudes toward learning that our entering freshmen exhibit. In fact university faculty can play an active role in helping high schools to prepare their students for college and university learning. It should be noted explicitly that the "high school method" of learning and the "college/university" method of learning are quite distinct. And students must be shown what the difference is and then taught how to pass from one mode of learning to the next. Certainly cooperation between high schools and colleges can play a major role in making such a program effective.

Recommendations

1. Mathematics departments, in partnership with other faculty responsible for teacher education, should establish strong links with local pre-college institutions. This partnership should involve activities that entail exchange of faculty and administrators.

   - The university math department should invite in-service teachers to visit the university (or college). Visiting teachers may meet mathematics faculty and participate informally and formally in mathematical activities ranging from chats about curricular priorities to seminars about selected topics in mathematics and/or innovative instructional strategies. Visiting teachers may also be involved in formal courses geared to teachers' needs and interests, and to allow teachers to teach some elementary college courses.

167

The street does not run only one way. University faculty may also visit colleges, junior colleges, and high schools and may participate in their activities. Of course there should be constant communication among the different faculties while these activities are taking place.

- University faculty can offer advice, in a collegial atmosphere, on how to connect mandated syllabi and tests to some serious, sound mathematics. Conversely, university faculty can learn more about what is going on in high schools and junior colleges, and may thereby become better prepared to teach the students who enter their universities.

- The cooperative efforts of university, college, high school, and other faculty can help to establish a mechanism for college undergraduates to become involved with secondary and middle schools and with classroom mathematics teachers. Such a program would pave the way for mathematics majors into pre-college teaching as a career.

- The university faculty member can convey to students, teachers, and educators that understanding mathematics involves sustained hard work and has its own rewards for those who responsibly undertake its study. Also, faculty at all levels and at every institution of learning should understand that students should be held to the high academic standards we believe they are capable of meeting.

2. The activities mentioned in item 1 above should be built-in features of universities and colleges made possible by university administrators and department chairs without the constant distractions of finding grant money. Logistics and solvency issues should be jointly addressed by schools, by their university partners, and cooperatively by their administrators.

3. Mathematicians should actively cooperate with teachers and educators in formulating and implementing federal, state and district policies. These should include policies related to

- mathematics teacher certification requirements;

- the continuing education of teachers (staff development), e.g. devising course and seminar offerings appropriate for pre- and in-service teachers;

- the writing of state frameworks;

- the writing of new curricula and related materials;

- constructively criticizing computer software and textbooks recommended for adoption.

In any state of the union, mathematics professors are the ultimate authorities on the subject area of mathematics. But they also have considerable expertise in areas of pedagogy. Exercised diplomatically, professors can use that expertise to help shape curricula, text choices, and even values and attitudes throughout the state. In particular, professors should take up the long-ignored gauntlet of developing a fruitful interaction with high school teachers and administrators.

Such an interaction can only help both styles of institution to serve each other more effectively, and will also result in meaningful exchanges both of information and of personnel.

Research Mathematicians and Research
in Mathematics Education

Members: Hyman Bass (Discussion Leader), Kenneth Bogart, Michael Fried, Cathy Kessell, Alfred Manaster, Steve Monk (Reporter), Blake Peterson

Questions for Day 1

- What interests you particularly about education research?
- What education research have you undertaken or made use of?
- What interactions between education researchers and research mathematicians have you taken part in or have you found particularly worthwhile?

Questions for Day 2

- What specific types of mathematics education research do you recommend as likely to be most useful and accessible to mathematics teachers in the classroom?
- What specific types of mathematics education research do you recommend as likely to be most useful and accessible to those designing mathematics curricula?

Mathematics education research is a field of inquiry into the nature of mathematical learning, as well as into the practice of mathematics teaching. It provides a foundation and methods for designing diverse teaching strategies and for studying their effects. The study of mathematical learning investigates the process by which students give meaning to and learn to employ mathematical ideas and practices, by making connections with and updating their prior knowledge and experience. Such investigations not only provide basic knowledge essential to the development of curricula and materials, but can significantly inform teaching practice as well.

It is vitally important for the mathematics research community to become better acquainted with the field of mathematics education research, with the many insights and perspectives on student learning it affords, and with the applications to practice it suggests. Actions should be taken in this direction by individuals as well as professional societies and institutions. As teachers and professionals, mathematics researchers should become acquainted and/or engaged with research in mathematics education in a number of ways. Among these are:

User of Information. Among the products of mathematics education research are "first order" informational studies about students, teaching, and learning that should be helpful to college teachers. These include studies of: (a) the effects on student attitudes and on student learning of work in collaborative groups; (b) transfer and non-transfer of knowledge between apparently related domains; (c) patterns of retention (or loss) of students' knowledge of mathematical ideas and procedures during their study in a particular course; and (d) relationships between student preparation for entry-level courses and course performance. Although such studies rarely give decisive answers to pedagogical questions, they provide a critical basis for the planning that takes place before teaching a course, as well as a framework for reflections on the effects of this teaching.

Teacher-Researcher. To improve or enrich one's teaching requires careful reflection on one's teaching experience. This process is made more disciplined and effective by the adoption of various techniques and strategies for gathering and objectively analyzing data. These include: keeping a teaching journal and collecting anonymous student writings about their mathematical ideas; carefully following the work and progress of particular students; systematically examining data with other teachers. Mathematics researchers should also consider the use of such techniques as means of pursuing deeper questions about their own teaching and their students' learning, as well as becoming more thoughtful and critical users of research in mathematics education.

Collaboration with Researchers in Mathematics Education. Mathematics education is fundamentally an interdisciplinary field of inquiry that draws on expertise in mathematics, psychology, and sociology. Many of its key ideas and methods originate in fields such as anthropology and philosophy. Since few individuals have strong backgrounds in such diverse fields, mathematics education depends on inter-disciplinary collaboration for its development. Mathematicians can make their most important contributions to the field of mathematics education through collaborations with scholars in this field who show a strong orientation toward mathematics. As is the case in other inter-disciplinary work involving mathematics, such collaborations are likely to be productive only when there is openness, honesty, and respect on all sides. This takes both care and thoughtfulness as to how mathematics might look to a professional in another field, who is interested in mathematics but not an expert in it, and how one who is steeped in mathematics can come to share in the point of view and expertise of a field quite different from mathematics.

Professional mathematics societies and institutions can play a number of roles with respect to helping individual mathematicians become involved with research in mathematics education in these ways. Among these are: publication of expository papers and annotated lists of references to studies that will be most immediately helpful to college mathematics teachers; invitations for invited addresses on research in mathematics education at society meetings and colloquia; inclusion in the programs of professional meetings of workshops and short courses on ways of carrying out teacher research; and active encouragement of collaboration between professional mathematicians and researchers in mathematics education. Above all, they should find ways to foster increased communication and collaboration between the communities of researchers in mathematics and researchers in mathematics education.

Contemporary Issues in Mathematics Education
MSRI Publications
Volume **36**, 1999

Appendix: Internet Resources in Mathematics Education

This brief list of Internet resources dealing with mathematics education is not intended to be comprehensive. Many other useful Web sites can be reached via links from the ones listed below.

Comprehensive Guides

- The Math Forum—Mathematics Education
  http://forum.swarthmore.edu/mathed
- The Mathematics Archives-Teaching Materials
  http://archives.math.utk.edu/teaching.html

Discussion Lists

- math-teach. Discusses mathematics education at the K-12 level. To subscribe, send a message to majordomo@forum.swarthmore.edu with the words 'subscribe math-teach' in the body.
- calc-reform. Discusses the teaching of calculus. To subscribe, send a message to majordomo@e-math.ams.org with the words 'subscribe calc-reform' in the body.
- mathedu. Discusses the teaching of mathematics beyond calculus. To subscribe, send a message to majordomo@warwick.ac.uk with the words 'subscribe mathedu' in the body.

Teacher Preparation

- NSF Collaboratives for Excellence in Teacher Preparation
  http://www.ehr.nsf.gov/ehr/due/awards/cetp/cetplist.htm
- Websites for NSF Collaboratives
  http://www.utep.edu/~pete, http://www.utep.edu/~pete/#cetp

Integration of Teaching and Research

- Undergraduate Research Opportunities
  http://forum.swarthmore.edu/students/college/urp
- NSF Recognition Awards for the Integration of Teaching and Research
  http://www.aas.duke.edu/nsf/other.html

Education Reform

- Mathematics and Education Reform Forum
  http://www.math.uic.edu/MER
- MAA committee on Calculus Reform and the First Two Years (CRAFTY)
  http://www.humboldt.edu/~mef2/crafty.shtml
- Calculus Reform
  http://forum.swarthmore.edu/mathed/calculus.reform.html
- Project Kaleidoscope
  http://www.pkal.org

Mathematics with Applications

- Connected Curriculum Project (Duke University)
  http://www.math.duke.edu/modules
  http://grandmac.calpoly.edu
  http://www.math.montana.edu/~frankw/ccp/home.htm
- Middle Atlantic Consortium for Mathematics and Its Applications Throughout the Curriculum (University of Pennsylvania)
  http://www.math.upenn.edu/~ugrad/macmatc.html

Integrated Science, Mathematics, and Engineering

- Science Core (Stanford University)
  http://scicore.stanford.edu
- Foundation Coalition (Rose-Hulman, Arizona State, and other institutions)
  http://fc.rose-hulman.edu, http://www.eas.asu.edu/~asufc